D0765347

REAPER FORCE

Dr Peter Lee is an academic at the University of Portsmouth and an expert on military drone operations. His specialist subject area is ethics and war, and he has been researching and writing about RAF Reaper operations since 2012. His interest in the human dimension of war first emerged at the bedsides of grievously injured and dying soldiers during the first five months of the 2003 Iraq War. At the time he was serving as an RAF chaplain at the British military hospital in Cyprus. Over time, the physical and emotional trauma he encountered there had a profound impact on him, and he eventually moved from chaplaincy to academia in 2008. He subsequently spent nine years as a lecturer in Air Power Studies at RAF College Cranwell.

Peter has been extensively involved in public and academic debate over the UK's use of the Reaper in remote air operations for several years. As a result he is regularly invited to lecture on different aspects of drone use to military, security, academic, political, religious, media and wider audiences. He has previously published *Blair's Just War: Iraq and the Illusion of Morality* (2012) and *Truth Wars: The Politics of Climate Change, Military Intervention and Financial Crisis* (2015).

REAPER FORCE

INSIDE BRITAIN'S DRONE WARS

DR PETER LEE

JB
JOHN BLAKE

Published by John Blake Publishing,
2.25 The Plaza,
535 Kings Road,
Chelsea Harbour,
London SW10 0SZ

www.johnblakebooks.com

www.facebook.com/johnblakebooks
twitter.com/jblakebooks

First published in hardback in 2018

ISBN: 978-1-78606-964-1

British Library Cataloguing-in-Publication Data:

A catalogue record for this book is available from the British Library.

Design by www.envydesign.co.uk

Printed and bound in Great Britain by Clays Ltd, Elcograf S.p.A

1 3 5 7 9 10 8 6 4 2

Papers used by John Blake Publishing are natural, recyclable products made from wood grown in sustainable forests. The manufacturing processes conform to the environmental regulations of the country of origin.

John Blake Publishing is an imprint of Bonnier Books UK
www.bonnierbooks.co.uk

For the members of the RAF Reaper Force
and their families

The views and opinions expressed in this book are those of the author alone and should not be taken to represent those of Her Majesty's Government, MOD, HM Armed Forces or any government agency.

CONTENTS

ACKNOWLEDGEMENTS

There are many people who have enabled this book to come to fruition, the majority of whom, for security reasons, cannot be named. I acknowledge you all here.

I begin with my amazing and supportive wife, Lorna, for her unwavering belief in this project and for her practical advice; my daughters, Samantha and Fiona, who are a continuous source of inspiration, laughter and pride; my mother, Sheila, and my late father, Peter, who taught me that life is an adventure to be lived to the full.

From the RAF, the current Air Officer Intelligence, Surveillance, Target Acquisition & Reconnaissance (ISTAR) Force Commander, RAF Waddington. He opened the gates to the Reaper Force for me with infectious enthusiasm, asking only for a fair telling of the Reaper story.

The RAF Waddington Station Commander.

Five current and previous Officers Commanding of 39 and XIII Squadrons, who have not only given me unparalleled, continued access to their facilities and people, but have actively promoted the idea of the Reaper story being told through the experiences of the operators. One Squadron Commander not only took me 'flying' on a day-long 'live' mission, but allowed me to try out the sensor operator's (SO's) controls to give me a real understanding of the equipment and the challenges of using it.

The current and previous Director of Defence Studies (RAF) have gone above and beyond in engaging with the Air Staff and the MoD to secure various permissions.

My three official reviewers have been with me from the outset – the Director of Defence Studies (RAF), a previous Officer Commanding 39 Squadron and Dr Peter Gray, Senior Fellow in Air Power Studies, University of Birmingham – and have also excelled in providing incisive critique and helpful additional detail throughout the book.

My unofficial team of reviewers expanded from an initial core of one active pilot, SO and mission intelligence coordinator (MIC) to include spouses and interested veterans of the Reaper Force I met along the way. They have added detail and context, and helped me with procedures, processes and terminology.

Jeff and Alicia Richard, whose son, US Marine Corporal Matthew T. Richard was killed in action in Afghanistan in 2011, and the US Marines, led by Squad Leader Sergeant Seth Hickman, who helped me to tell Matthew's story.

Andrew Lownie, my literary agent, who did so much to help me secure a publisher.

My friends and colleagues at the University of Portsmouth who have provided practical and moral support throughout the project.

My friends and colleagues across the academic 'drone' community, and those from the anti-drone activist community, with whom I have discussed, reasoned and argued over several years, and who have shaped my thinking in the process.

The past and present members of the RAF Reaper Force, and their spouses and partners, who have welcomed me into their community, laughed with me, shed tears with me and entrusted me with their most profound experiences. I promised each of you that I would be honest, accurate and fair, and have strived to match that undertaking at every stage.

Andrew Jeffrey, Joel Anthony, Hayden Hunt and James Robertson for contributing photographs which, with others, are credited in the captions to the photos in the plates section.

Toby Buchan, my editor and great source of encouragement and guidance.

I give you each my profound thanks. This has been the most exciting, moving, challenging and fulfilling project of my life. All of the credit belongs to you and any failings in the pages that follow are mine.

TERMS AND ABBREVIATIONS

9-LINE A series of nine confirmations and checks, and the legal approval, that make up the authorisation to carry out a weapon strike, and includes the location and identity of the target, and the weapon choice

AIRFRAME The structural elements of an aircraft, including the fuselage, wings and undercarriage, but excluding the engine. RAF personnel sometimes use 'airframe' interchangeably with 'aircraft'

AUTH Authorising Officer. Supervises Reaper crews and operations from the Squadron Operations Room

BDA Battle damage assessment

BUDDY LASE When two aircraft and crews work together to strike a target – one aircraft fires a laser-guided missile or bomb, while a second uses its laser-guidance system to ensure that the weapon hits the target that the laser is 'lighting up'

CAO Casualty Assistance Officer

CAOC Combined Air Operations Centre

CASEVAC Casualty evacuation

CAT Flying category (e.g. Combat Ready)

CDE Collateral damage estimate

CIVCAS Civilian casualties

CNO Casualty Notification Officer

CRM Crew resource management. The sharing out of responsibilities and tasks between the three crew members

COMMS Communications. Commonly by radio, electronic signal or secure military internet

CPL Corporal

DICKING The observation of military personnel and movements by, usually, low-level, unarmed enemy called 'dickers' (in Afghanistan they would often be young boys). The term was used in Northern Ireland and referred to junior, or would-be, paramilitary members

EOD Explosive Ordinance Disposal

FAST JET A generic British military term that refers to high-speed, jet-propelled, air-to-air fighter aircraft or fighter-bomber aircraft.

FMV Full motion video

FOB Forward Operating Base

GBU Guided Bomb Unit. The Reaper carries two 500lb GBU-12 laser-guided bombs

GCAS Ground close air support

GCS Ground Control Station

HME Home-made explosive

HVT High-value target. For example, an important Taliban figure

IED Improvised explosive device

IN THE BOX In the Ground Control Station, a shipping container-sized metal box

IS Islamic State

ISAF International Security Assistance Force, the NATO-led mission in Afghanistan

ISR Intelligence, surveillance and reconnaissance

ISTAR Intelligence, surveillance, target acquisition and reconnaissance

JTAC The person who directs offensive air operations, including close air support, typically from a forward position on the ground or from a Combined Air Operations Centre

KINETIC STRIKE A missile or bomb strike

LASING Using a laser beam to mark or 'light up' a target for a laser-guided missile or bomb strike

LCPL Lance Corporal

LRE Launch and Recovery Element – the crew based near the operating area who launch the Reaper and land it at the end of the mission, by remote control. In between take-off and landing, control is transferred electronically so that it is then piloted via satellite link by someone in the UK or USA

MATCH EYES To train the Reaper camera on a target or area that is already being watched by another surveillance aircraft

MCE Mission Control Element

MEDEVAC Medical evacuation

MIC Mission Intelligence Coordinator

NOTAM Notice to Airmen – a standard, internationally recognised method of warning aircrew of mandatory areas to avoid and potential risks in the area where they are planning to fly

OFF DRY To pull out of a bomb or missile run without releasing a weapon

OVERWATCH To fly above friendly forces on the ground, looking out for threats and using weapons to protect those forces if necessary. Ground forces can also conduct overwatch, with one group of soldiers or marines observing as their colleagues manoeuvre into position

PAX Passengers

PID Positive identification, usually of a combatant target

PLATFORM An aircraft, which can be manned or remotely piloted, plus all of its capabilities

POL Pattern of life, built up through intelligence gathering and surveillance

PTSD Post-traumatic stress disorder

QWI Qualified Weapons Instructor

RAR Remedial Action Report

RCDFCIED Remote Control Directional Fragmentation Charge Improvised Explosive Device

RCH Red Card Holder – the UK's RCH holds and provides operational and legal authority for a weapon strike that has been delegated from the Secretary of State for Defence, via the Air Component Commander in the CAOC

ROE Rules of Engagement – the legal framework that dictates when, where and how lethal force can and cannot be used

RPAS(P) Remotely Piloted Aircraft System (Pilot) – the official designation of someone recruited and trained to pilot an RAF Reaper, who was not previously a pilot on another type of aircraft

SAFETY OBSERVER An experienced Reaper crew member, or instructor, who is brought into the GCS in the build-up to a missile or bomb strike

SANGAR A fortified position

SAR Synthetic aperture radar

SMIC Senior Mission Intelligence Coordinator. Supervises Reaper missions alongside the Authorising Officer in the Squadron Operations Room

SO Sensor Operator, who controls the various cameras and other sensing equipment, and who laser-guides missiles and bombs onto targets

TECHNICAL Flat-bed truck with a large weapon bolted or welded onto the back; used as highly mobile light artillery

TIC Troops In Contact

TRIM Trauma risk management – peer-to-peer psychological support system intended to reduce the impacts of traumatic events

USAF United States Air Force

VBIED Vehicle-borne improvised explosive device

VOIED Victim-operated improvised explosive device

WEAPON OFF THE RAIL When a missile has been fired or a bomb dropped from an aircraft

PREFACE

It was a close friend who prompted my drones research in 2011 by introducing me to someone in the RAF Reaper community. I thank him and curse him every day in equal measure for what he started.

I have a love-hate relationship with all things war and military-related. My own experience of war took place well away from where the bullets are fired and the bombs go off. I was an RAF chaplain for seven years, from 2001 to 2008. I have never been on a front line, never been shot at and never fired a weapon in anger against an enemy. The closest I ever came was during a year in the Falkland Islands, where I got to walk old battlefields with veterans who fought in 1982 and who, in 2004 and 2005, were still wrestling with the mental legacy. My involvement in the 2003 Iraq War was as a chaplain at a British military hospital in Akrotiri, Cyprus, over a five-month period. There, on a daily basis, I witnessed the brutality of war in the broken, wounded and maimed bodies of (mainly) young soldiers and the occasional Iraqi civilian. Over time, those encounters took their own psychological and spiritual toll on me.

Until I started my research for this book I had never watched someone being killed. I mean *actually* killed, in real time, as opposed to seeing a YouTube video or some fictionalised Hollywood scene where the actors take off their make-up at the end of the day and go home. No, my experience of war has been routine, mundane and focused on the human cost – sometimes painfully so – not on the military or political aspects.

I was initially reluctant to pursue the project. I had spent several happy years away from the physical and mental traumas of war, teaching the ethical and political aspects of armed conflict and air power from the safety of a classroom. To delve into the professional and personal lives, and the experiences of the people who kill using remotely piloted Reaper aircraft – commonly referred to as 'drones' outside the RAF – in a new form of war from a distance, would be interesting. But it would also require me to return to an inner place I did not want to revisit. Then there was the practical difficulty of gaining the necessary level of access to one of the world's most guarded and classified military communities.

My questions were endless. Would it even be possible? If I did manage to gain access would the Reaper operators be willing to talk to me? Why should they? Would their spouses and partners want to share their experiences of living with those who brought the mental images of distant killing home with them? It would be much easier by far to stay in the classroom, to stick with the theory rather than immerse myself in the world of the Reaper operations and those who carry them out.

Just as I was deciding to ignore the idea, fate intervened. I spent several hours with a Reaper pilot, John, and his wife, Kim, who spoke candidly about their experience of fighting wars on two fronts. He was fighting Islamic State (IS) jihadists at work and she was fighting cancer at home. As they described life in a strange in-between world that spanned killing and healing, I knew I had to try to write this book. Despite the apparent lack of physical danger John faced in piloting a Reaper over distant continents, this new form of war clearly carries different risks. The jeopardy is not from Taliban or IS bullets and bombs but is, instead, psychological and relational.

I started the process of trying to gain access to conduct the necessary research. My approach to the RAF Director of Defence Studies in early 2015 was simple and emerged from two historically linked questions: How much would people today – 100 years on – like to know more about the lives of the pioneering aircrew of the First World War? Will people today and in the future be similarly interested in the lives and experiences of the first generation of Reaper operators?

We are aware of what those First World War pilots and rear crew did in the air because of the flying log books and official records that survived them. Yet little is understood about how they felt about their experiences and day-to-day lives, partly because so few of them kept diaries, partly because the notion of publicly sharing personal feelings and sensitivities would have been culturally alien and partly because so few of them survived aerial combat to tell their stories. I wanted to make sure that such information about today's Reaper crews was fully recorded. There is considerable public, academic and political interest in the first generation of Reaper operators, fuelled by the fact that their lives are even more secretive and less accessible than those of the First World War aircrews.

Within days of my approach, an informal, positive response came from the Air Staff, the senior officers who run the RAF, followed by several months of administrative action to bring the project idea to life. Eventually a letter of authority was granted but one final hurdle remained: the RAF, MoD and my university research ethics committees. In theory, ethics committees exist to promote quality research and protect participants. In practice, some of them can seem more like research prevention committees. Ultimately, however, with safeguards in place to ensure personnel and operational security, most notably through anonymity, the final research permissions were granted in June 2016.[1] I was committed. All it needed was for the Reaper operators to be committed or all this would have been for nothing.

In July 2016, with my final clearances in hand, I travelled to 39 Squadron (RAF) at Creech Air Force Base, Nevada, for a week. It was a step into the unknown. I had a couple of contacts on the squadron who would host me during my research visit and, hopefully, encourage others to take part and be interviewed. And take part they did. I ran out of time in Nevada to interview all the operators and their partners who volunteered during the course of my stay. The next month I found myself embedded for two weeks with XIII Squadron at RAF Waddington in Lincolnshire. Again, I eventually had to return to my lecturing without completing all the interviews I was offered: they would continue sporadically through 2017 and into 2018, with further visits to both XIII Squadron and 39 Squadron. Eventually, I would formally interview 90 members of the Reaper community: 45 currently serving personnel, 21 former Reaper crew members, and 24 spouses and partners.

This book is neither a systematic treatment of all aspects of the RAF Reaper Force, nor is it a proportionate representation

1 'Royal Air Force Reaper: 21st Century Air Warfare from the Operators' Perspective', University of Portsmouth Research Ethics Committee Protocol E365, approved 21 October 2015; Ministry of Defence Research Ethics Committee Protocol 707/MODREC/15, approved 1 July 2016.

of everything that it has done over the past ten years. Page after page describing many hundreds of similar attacks or countless hours of dull reconnaissance activities would be boring. I have therefore focused on a series of events and accounts that provide important snapshots in time, each of which says something unique and interesting about the people who fly the aircraft and the families that support them, and the development of the Reaper Force.

It is also not an official history, although it is historical all the same with accounts of real people and real events at particular times and places. At least two events that are covered in depth were the subject of major international press stories. Given the highly detailed and intimate approach I adopt throughout the book, I did not have room to include all the information I was given by the many contributors.

The book has been researched and written against the backdrop of ongoing terrorist threats against military personnel in general and Reaper personnel in particular.[2] Consequently, the key requirement for my gaining access to the Reaper crew members and their families is that their identities and personal security should be protected. To that end, pseudonyms have been used throughout. I do not know the names of everyone, past and present, on the Reaper Force. If my chosen pseudonyms match the names of real people in the Reaper community I have not met, it is entirely coincidental.

The human memory is an imperfect tool, but even those imperfections have an authenticity of their own when they are caused by the intensity or trauma of war. To ensure accuracy of tone and content, and to meet my personal writing ethic of

---

2 *The Times*, 1 May 2016, 'Isis hackers publish hitlist of drone pilots', https://www.thetimes.co.uk/article/isis-hackers-publish-hitlist-of-drone-pilots-xz59sq5bb, accessed 15 February 2018.

'honest, accurate and fair', I have gone back and checked all the information with the contributors. Far from being asked to withhold details, I was delighted to receive countless snippets of additional detail and insight.

All credit for the success of this project goes to those who made it possible and those who have shared their most personal experiences. I can only share with you, the reader, what others have been willing to share with me. Any deficiencies or inaccuracies are my responsibility.

PETER LEE

INTRODUCTION

'I DROPPED MY SON AT SCHOOL IN THE MORNING, CONTINUED ON TO WORK AND, WITHIN A COUPLE OF HOURS, KILLED TWO MEN. I WENT HOME LATER THAT DAY TO BE GREETED BY MY SON WITH A CHEERY, "HOW WAS YOUR DAY?" DO YOU LIE TO PROTECT HIM OR DO YOU TELL THE TRUTH?'

JAY, REAPER PILOT

This is a book about the unknown community of the RAF Reaper Force. It is a group that embodies a series of contradictions: aircrew who never leave the ground, who are unseen but regularly in the news and who operate at the cutting edge of technology yet rely on the basic roles of air power – surveillance and attack – that have existed for more than a century.

The biggest contradictions, however, surround the aircraft they fly, remotely via satellite links, from distant continents: the MQ-9 Reaper. For many, perhaps most, people outside of the military, the Reaper is a drone. The word 'drone' implies that they are autonomous, self-thinking, emotionless robots but this overlooks the vast technical infrastructure and hundreds of people needed to operate the Reaper squadrons, and ignores the three crew members that fly each aircraft from a Ground Control Station (GCS): the pilot, SO and MIC. Furthermore, 'drone' becomes almost meaningless because it puts the Reaper in the same category as small hobby quadcopters. These hobby drones typically weigh less than 20lbs, measure 30–60 centimetres in diameter and typically stay airborne for less than an hour. They must be flown 'line of sight' (that is, you have to be able to see them with the naked eye), and fly no higher than 400 feet.

In contrast, the Reaper is a fully functioning aircraft with a 60-foot wing span that is piloted remotely from a GCS far away. It can carry four 100lb Hellfire missiles and two 500lb laser-guided bombs, operates at 20,000 feet, and can stay airborne for between 12 and 20 hours depending on its weapon load. To fly a Reaper the pilot has to pass aviation exams and (s)he must follow the aviation rules and laws that pilots of manned military aircraft must follow. Practically and legally, the International Civil Aviation Organisation recognises the Reaper as a remotely piloted aircraft (RPA). When combined with all the elements that make it work, like satellite communication, computer links, crew and infrastructure, it is formally known as a remotely piloted aircraft system (RPAS).[3] Similarly, RAF personnel and RAF air power doctrine refer to Reaper, aircraft, RPA or RPAS. Therefore, for the sake of accuracy and authenticity, unless the context calls for the use of the term 'drone', such as my entry into the Reaper world in Chapter 1, the remainder of the book will

---

3 RPAS is the legally designated descriptor used by the International Civil Aviation Organisation. See http://www.icao.int/Meetings/anconf12/IPs/ANConf.12.IP.30.4.2.en.pdf, accessed 23 July 2018.

also refer to Reaper, aircraft, RPA or RPAS as I take the reader inside what is more widely referred to as Britain's drone wars.

What became known as the 'drone wars' started with America's use of the MQ-1 Predator and the MQ-9 Reaper in the years after 9/11, which has attracted significant criticism due to the numbers of civilians that have been killed in the course of the USA's War on Terror. Millions of words have been written about CIA activities involving the Predator and Reaper that have taken place as far afield as Pakistan, Yemen and Somalia, amongst the best of which is Chris Woods's *Sudden Justice: America's Secret Drone Wars*. A popular vocabulary of war has been spawned to describe the CIA's use of Predator and Reaper, using the term 'drone', as well as a vocabulary to describe those who fire the missiles: 'Playstation killers'[4] who are said to be detached, disengaged, remote and emotionally disconnected. Many of those assumptions about America's use of Predator and Reaper 'drones' have been applied to the RAF Reaper Force. The chapters to follow will challenge these stereotypes and assumptions.

I had my first conversations with a few British Reaper pilots and SOs in late 2011 and 2012. I had been out of the RAF for several years by then and approached them with a negative attitude towards their work, which reflected the general tone of media comment at the time. But, once I got talking to them, I found an intense mental and emotional engagement with what they did and with the people they targeted. The quote at the start of this chapter was merely the starting point for deeper and more insightful conversations in subsequent years.

A major source of frustration for the British Reaper personnel has been, and still remains, that the actions of the RAF's two Reaper squadrons have been conflated by the media and anti-

4 Cole, C., Dobbing, M. and Hailwood, A., *Convenient Killing: Armed Drones and the 'Playstation' Mentality* (Oxford: Fellowship of Reconciliation, 2010).

drone lobby with the CIA's use of the Reaper and the Predator. (I have spoken informally to several USAF Reaper and Predator pilots who did not like being linked with the actions of the CIA. Specifically, they did not want to be associated with the large numbers of civilian deaths attributed to the CIA.[5]) So, I found it difficult to reconcile the experiences of the RAF Reaper personnel that I was beginning to encounter with some of the more extreme claims that were being made about 'drone' operators. For example:

> He is a drone 'pilot'. He and his kind have redefined the words 'coward', 'terrorist' and 'sociopath'. He is the new face of American warfare. He is a government trained and equipped serial killer. But unlike Ted Bundy or John Gacy, he does not have to worry about getting caught. It is his job… A CIA strike on a madrassa or religious school in 2006 killed up to 69 children, among 80 civilians.[6]

There was something akin to an obsession with zero CIVCAS (civilian casualties) among the British crews I spoke to and later observed in action. The attitude was influenced by RAF civilian casualties from an incident in 2011 (see Chapter 5). I was a King's College London lecturer specialising in the ethics of war and air power in 2011 and 2012. At the time, the same weapon system – the Reaper – was reported as producing high numbers of civilian casualties by the CIA, while the RAF was

5 Philip Alston, 28 May 2010, Report of the Special Rapporteur on extrajudicial, summary or arbitrary executions, Study on Targeted Killings, Human Rights Council, UN Doc. A/HRC/14/24/Add.6, http://www2.ohchr.org/english/bodies/hrcouncil/docs/14session/A.HRC.14.24.Add6.pdf, accessed 5 February 2018.

6 Vic Pittman, *Salem News*, 18 April 2013, 'Cowardice Redefined, The New Face of American Serial Killers', http://www.salem-news.com/articles/april182013/american-killers-vp.php, accessed 10 February 2018.

reporting one incident. My academic background in the ethics of war, and years of lecturing on military air power, told me that the difference in casualty numbers had to come down to the contrasting policies of different governments and the Rules of Engagement (RoE) that they imposed on their respective Reaper Forces.

A scalpel in the hands of a skilled surgeon is used with precision and purpose, but still damages healthy tissue; a scalpel used without skill, or used by a torturer, can disfigure faces and bodies. The Reaper with its 100lb Hellfire missiles can be unbelievably accurate in, say, hitting a moving vehicle or a single individual. However, I do not want to draw too strong a comparison between the relative accuracy of a scalpel and a Hellfire missile. A 'surgical' missile strike can be very accurate compared to the use of a 'dumb', unguided bomb, but it is obviously not the same degree of precision as a surgeon achieves in a hospital. A small missile is still a missile: fire it into a dense crowd of civilians and many of them will be killed and wounded. A 500lb bomb will make a proportionately bigger explosion with the potential to kill more people, combatants or non-combatants.

The key question is this. How many civilian deaths will a government allow its armed forces to inflict in the pursuit of the government's aims? Bluntly, the US considers itself to be at war and has been since 9/11, while the UK has chosen to participate in several military operations during that same period. These political differences have dictated the degree of force that successive US and UK governments have been willing to allow their Reaper Forces to use. Many individuals and organisations either do not know or understand these subtle differences, or they have ignored them for the purposes of the arguments they want to make. For example, the charity War on Want stated:

> Drones are indiscriminate weapons of war that have been responsible for thousands of civilian deaths. Rather than expanding the UK's arsenal, drones should be banned, just as landmines and cluster munitions were banned. Now is the time to stop the rise of drone warfare – before it is too late.[7]

This statement links 'thousands of civilian deaths' with the UK's drones – the Reaper, in other words – despite evidence including the UN's 2010 report on 'extrajudicial, summary or arbitrary executions' that does not even mention the UK or the RAF Reaper in its criticisms of drone use.

I have deliberately laboured these political and technical issues here because they provide the background to both my research and to the operations carried out by the RAF Reaper Force. They should also be borne in mind during the chapters to follow. From here on, however, my focus is on the pilots, SOs and MICs, past and present, who conduct the UK's Reaper operations. They do not work in isolation, however, but are the most visible part of a vast and complex system. It takes an extensive array of people and skills, across several countries, to get a Reaper airborne and to enable it to function across continents: engineers of different types, communications specialists, computer programmers, operations support personnel, weapons technicians, armourers, intelligence gatherers, imagery analysts, air traffic controllers, aerospace battle managers, Joint Terminal Attack Controllers (JTACs), lawyers, logisticians, flying programme administrators and many more besides.

Crucial though, and often overlooked in conventional books about war or air power, are the spouses, partners, families and

7 War on Want, 'Killer Drones', https://waronwant.org/killerdrones, accessed 10 February 2018.

friends that send the Reaper crews off to war every day. They live with the pressures, the fatigue and the emotional overflow from the daily operations conducted by the crew members. Many also spend a large proportion of their time as, effectively, single parents. It is not possible to write about the members of the Reaper Force in isolation without showing how their work impinges on family life, and vice versa.

And now, the structure of the book, which seeks to immerse the reader in the lives of the Reaper operators, from the claustrophobic, fast-moving, life-and-death decisions in the GCS to the human cost, the moral dilemmas, and the triumphs and failures that they experience. There is no strict chronological sequence, as RAF Reaper operations in Afghanistan, Iraq and Syria are explored. Depending on who is speaking, and when, several names are used for the group who sought to establish a caliphate across Syria and Iraq: 'ISIS' (Islamic State of Iraq and Syria), 'ISIL' (Islamic State of Iraq and the Levant), 'Dacsh' and 'IS' (the self-proclaimed Islamic State). I use 'IS' for its brevity, while remembering its apocalyptic ambition and vision and, as a United Nations Human Rights report stated, for 'imposing their radical ideologies on the civilian population'.[8] This radical ideology is enforced by extreme violence, enslavement, rape, murder, torture and more.

In the first three chapters, I take the reader into 'Reaper world' through my own experiences as I set out on my research. Chapter 1 follows the journey from Las Vegas to Creech Air Force Base and into 39 Squadron, concluding with the pre-flight briefing for the day's mission. Chapters 2 and 3 are spent in the GCS with two different crews on successive days. The first day captures

8 Pinheiro, P.S., 18 March 2014. Statement by the Chair of the Independent International Commission of Inquiry on the Syrian Arab Republic. United Nations Human Rights, http://www.ohchr.org/EN/NewsEvents/Pages/DisplayNews.aspx?NewsID=14397&Lang ID=E, accessed 12 February 2018.

hour after hour of surveillance activities, while the second day sees two missile strikes against IS jihadists. Chapter 4 is the most historically focused of the book, providing an insight into the origins of the UK Reaper Force through the eyes and experiences of several pioneers who developed their understanding while serving on exchange with the USAF Predator fleet.

Chapters 5-9 are all highly detailed and operationally focused, highlighting specific individuals or experiences. Chapter 5 recounts the 2011 civilian casualty incident through the eyes of the MIC involved, going on to reflect on living with what happened that day. Chapter 6 challenges many assumptions about gender and war as 'Tara' flies Reaper operations, including her employment of lethal weapons, through to advanced pregnancy. Chapter 7 provides a major 'What if?' moment for a crew whose target fixation could potentially have had a disastrous outcome, then follows the retraining process that brings them back to full capability again. Chapter 8 re-lives the moral dilemmas of one Reaper SO who struggled with killing, and controlling missiles onto human targets. Then Chapter 9, Happy Boxing Day, proves to be anything but as a crew spends Christmas night 2014 helplessly observing a series of horrors on the ground in Iraq, whilst being unable to intervene to protect the 'friendlies' below.

The final four chapters are more reflective. Chapter 10 brings together the thoughts and experiences of a range of Reaper crew members in their own words. Chapter 11 provides a moment-by-moment account of one of the most famous RAF Reaper missile shots from those involved. The shot disrupted a public execution and the video was released by the MoD at the same time as it was announced that there would be no medallic recognition for the Reaper Force personnel for their fight against IS. The penultimate chapter gives voice to several spouses and partners who speak about their lives with those who go to war every day, and the challenges they face. The final chapter combines my own thoughts on what I have seen and experienced, with reflections from Reaper operators on the personal legacies of having been at war, continuously, for up to seven years and more. One additional, unplanned chapter appears as an Epilogue and captures a glimpse of the human cost of war. It started as a brief, personal account from a Reaper MIC who witnessed the death of a US Marine, Corporal Matthew Richard, in Afghanistan, and still remembers it every day. The story evolved over several months as I got to know Cpl Richard's parents, his Squad Leader and five fellow squad members, who all shared their perspectives on what happened on 9 June 2011.

CHAPTER 1

INTO THE DRONE LAIR

'YOU'RE FLYING WITH ME TODAY'

SQUADRON BOSS

It is an odd feeling, knowing that I am about to watch someone be killed. Perhaps not today, and perhaps not even tomorrow, but almost certainly before the week is over I will see someone's life ended before my eyes. Deliberately, precisely and with extreme prejudice, using a missile or bomb from an RAF MQ-9 Reaper – a drone, in popular terminology, or an RPAS to use its more technical term. The very possibility transports me to another time and place that I try not to think about too often. To a military hospital in Cyprus in 2003 where I spent five months of the Iraq War at the bedsides of the wounded, injured and dying. Maybe I haven't thought this through properly. But here I am, not in Cyprus… in Vegas. And not to gamble. (No way. The house always wins.) My cheap casino room is merely a base from where I will venture into one of the most secretive communities on Earth.

An hour's drive from the epicentre of Vegas's hedonism stands Creech Air Force Base, home to an array of USAF capabilities, the most famous – or infamous, depending on your opinion – of which are the MQ-1 Predator and MQ-9 Reaper. It is also one of the two places from where the UK operates its own Reapers, the other being RAF Waddington. I am about to spend several days alongside the RAF Reaper crews of 39 Squadron, with a behind-the-scenes view of the war they are waging against IS in Syria and Iraq. I cannot dignify the jihadists' self-styled use of the bogus name 'Islamic State', and I'm not sure how neutral I will be as I watch events unfold. Any group that kidnaps, sells and rapes thousands of young girls, and murders Muslims and others in pursuit of its aims, does not have the hallmarks of Islam or of statehood.

The early morning traffic is accumulating rapidly as my condescending satnav guides me briefly along the Las Vegas strip and then away on a circuitous route to US95. Highway 95 takes me northwest out of Vegas towards Creech where 39 Squadron has been based since its re-formation as a Reaper Squadron in 2007.

As the Vegas suburbs give way to desert the transition comes quickly. The last signs of civilization – if that is the right word – are the power cables above the highway. High- and low-voltage cables: the twenty-four-hour pulse that keeps the city alive. No electricity, no lights; no electricity, no pumped water; no electricity, no pinging slot machines.

There are no signs of life as far as the eye can see. The car thermometer says it is 43^{0}C (109^{0}F) outside and the sky is a rich morning blue, edged with paler shades around the horizon. No clouds, no hope of rain. The road is arrow straight for miles ahead and a constant shimmer maintains its distance half a mile in front of me as the sun works its magic on the tarmac. In the mirror an identical shimmer follows me a half mile behind. Las Vegas begins to disappear. The ultimate illusion: making a whole city – this city in particular – disappear in its own mirage.

A few miles to the left and right of the road stand parallel rows of mountains. Perversely, given the desert heat, I pass a road sign that points to a distant, elevated ski resort. As my eyes follow the direction of the sign the mountains become craggy, foreboding. Their rock strata emerge from the ground at shifting angles. Immediately to my left a peak rises steeply, while for several peaks further on the angle to the land below is much shallower. I wonder about the forces that can shape trillions of tons of rock in that way over millions of years, or over just six days, according to the sign I saw earlier.

To my right a twin mountain range stands over the desert, except this time the strata are contoured – waves of rock rather than hard, straight lines. Sandstone colours are packed between the darker, harder layers. Between both sets of mountains lie desert and scrub, bisected by black tarmac. A sign to the right points towards a tribal Visitor Centre on one of thirty-two Nevada Indian reservations and colonies spread across the state.[9]

An upward glance spots criss-crossing vapour trails high overhead, probably commercial airliners. Much lower and far more interesting are two pairs of fighter aircraft – they look like F-16s and F-22s – flying several miles apart, one pair in close formation, the other combat formation. From their trajectory they are heading back to Nellis Air Force Base from the Nevada Training Range. I recall watching *Top Gun* when it first came out in 1986 and I wonder which of these pairs 'won' their aerial dance of death. I like to imagine that a couple of grizzled, middle-aged dudes in the F-16s whipped a pair of young guns in the F-22s.

9 For further information see http://www.nevadaindianterritory.com/, accessed 10 June 2018.

In reality, the F-22s would have released a couple of over-the-horizon missiles and ended the argument before it began. The jets remind me that this trip is about flying, not sightseeing. But the aircraft I am interested in are flying over a different desert landscape on another continent.

This is the road that thousands of Reaper and Predator crews have used over the years on their daily commute to Creech. These have been mainly USAF personnel. However, for over a decade a fair smattering of British crews have joined their number, and it is their stories I am here for.

I wonder how many of the Predator or Reaper pilots have watched fighters like these, masters of the sky, and envied the pilots who do their flying up there? How many of them have been there and done that, and are now happy to swap fighter cockpits for static cockpits: metal containers firmly anchored to the ground? Another question springs to mind from the numerous debates, discussions, media events and conferences on drones I have taken part in. How many people have no idea that Reaper and Predator 'drones' are every bit as piloted as the F-16 fast jets now rapidly disappearing in the distance? The only real difference is that instead of several feet between the fighter pilot's controls and the aircraft's engine and wing surfaces, signals from the Reaper pilot's controls travel several thousand miles via satellite.

I put on the radio and find a local station. Someone is beyond excited about today's sunny weather in Las Vegas, which makes me wonder if a corresponding radio presenter is equally excited about another day of snow in Alaska, thereby keeping climate karma in balance.

Then the morning show presenter introduces Coolio and I chuckle to myself. A distinctive synth riff sits over a groaning bass line and leads into the only lyric of his that I know: 'As I walk through the valley of the shadow of death, I take a look at my life and realise there's nothin' left.' From Psalm 23. Here I am driving through a valley of death, but now with biblical allusions to add to my self-induced mind games. Thanks Coolio.

Every military funeral I have conducted or attended has featured Psalm 23 and I visualise them in a mental torrent. Countless war memorial services flash through my mind until I settle on one I conducted in 2005 in the Falkland Islands for a small group of Welsh Guards, veterans of the 1982 war. One of them read those later words from Psalm 23, 'Surely goodness and mercy shall follow me all the days of my life.' I was angry then and I begin to feel angry now. The only thing that had followed those particular soldiers throughout their lives was scarring from third degree burns and PTSD.

I try to focus on the road. I cannot see how this particular drive will distract the Reaper crews from their thoughts of the death that they regularly consider and occasionally deliver. Then slowly, finally, Creech starts to sneak into view in the distance. First, some small buildings that can barely be seen through the heat haze rising up from the road. Then, gradually, expanses of concrete, roads and runway come into view. Runway lights on their intricate frames point the way to the landing threshold. As I get closer, more and more buildings appear. And then a green sign against the mountain backdrop: 'Creech AFB'.

I take a few moments to reflect on how long and difficult it has been to get this far. Not to make the simple drive out from Las Vegas. Rather, the process that started with a simple phone call to the RAF's Director of Defence Studies almost a year and a half ago. To his credit he didn't put the phone down on me, though if he knew what lay ahead he may well have done. What I thought I was requesting was access to some interesting people to capture what I hoped would be some interesting stories. What he heard was someone asking to access one of the most secretive and controversial programmes in the armed forces, in the hope that enough of these over-busy men and women would be willing to talk to him in enough numbers to fill a book.

I'm sure I still don't know the full extent of the work he had to do on my behalf; how many phone calls and emails it took. For his own security I can't even give him a name check. Yet here I am, nearing Creech, with a wad of clearances, approvals and passes stacked on the front seat.

I swing off the main carriageway onto a narrower road that winds round to the checkpoint at the entrance to the base. The speed limit drops to 15mph and suddenly everything looks more ominous with 'Stay Out' signs and razor wire topping the fences. I queue behind two cars and watch the guards, attentive and serious as they check vehicles, identification and documentation. There are no cursory glances. I have entered military bases on several continents and am yet to meet any guard who takes their duty more seriously than an American soldier under orders to keep a place safe.

Both guards are decked out in the full Robocop: body armour, two-way radio, pistol, rifle, fingerless gloves with armoured knuckles, reflective sunglasses. They look like they could storm a bridgehead or defend an isolated outpost in Helmand Province, Afghanistan. Maybe they already have.

Guard 1 does not smile as she bends to look in the first car and talk to the driver. Guard 2 has her head on swivel mode and almost seems to be in a slow-moving dance, her partner being the rifle she holds across her torso. I wonder if the safety catch is on. She looks in the back of the first car, all the while glancing up at me and the vehicle between us. Are the RayBans to keep out the sun or to keep a psychological barrier between the watcher and the watched? Perhaps both. It takes a lot of self-discipline to pay this much attention to detail in stultifying heat.

Car 1 is waved through and Car 2 gets the same treatment. The smiles and relaxed banter of countless guards I have met at the gates of many British military bases are conspicuous by their absence. I have been warned not to try to make small talk. Jokes and witticisms are out, while sudden movements would make things go loud and ugly. I glance once more at the passenger seat where my passport and documents are neatly, obsessively neatly, clipped together. Two years of preparation, worrying and hoping have come down to this: whether a private soldier recognises and accepts the security certificate that was printed, signed and stamped by an RAF police sergeant a continent away.

I am motioned forward. Still no smile. I recognise a distinctly Scottish surname velcroed to the body armour of Guard 1.

'Identification and purpose of visit?'

'I am visiting 39 Squadron for a few days.' I hand over my passport and papers, and she leans away to pick up a clipboard from the adjacent booth. I have arrived at precisely my allotted arrival time.

The documents are scrutinised for errors or flaws and checked against whatever is written on her clipboard. Without looking up, she asks, 'You Scottish?' If there is an ounce of Scottish blood in her ancestry things might get a bit more relaxed.

'I'm from Dunfermline near Edinburgh, originally.' I could see her face soften. The Celtic connection was made.

She replies, 'My grandmother is from Edinburgh, from a place called Pilton. I've always wanted to visit.'

Take your gun with you if you visit Pilton, is what I was thinking. What I said was, 'I hope you get the chance – go in the summer time.' It would all be plain sailing now.

'Where is your escort?'

'Escort?'

'Yes, you need to be escorted at all times.' My plain sailing ship just sailed. I tried not to panic.

'I have some other printed instructions in my bag. Can I check them?' No unauthorised movements – she has a gun and a bloodline from Pilton. She nods.

I start pulling out every bit of paper related to this project, and there are a lot of them. Out of the corner of my eye I spot a car pull up on the inside of the checkpoint. Someone in a flying suit steps out of the car, and the lack of weapons and military bearing tell me right away that he is from the RAF. My escort?

Guard 1 walks over to him, exchanges a few words and checks his ID. She fills in a pass which, smiling, she hands to me with my other paperwork. 'Welcome to Creech Air Force Base.'

As he jumps back in his car, a series of incomprehensible hand signals from my escort suggest that I should follow him. He holds up his palm with fingers and thumb splayed wide. Is he indicating five miles or five minutes? I wave, nod and move before I get left behind.

There is little chance of getting lost at this stage. On either side of the sand-strewn road stands an honour guard of 2ft high boulders that would wreck the underside of a truck if it decided to take a detour. In the distance more buildings come into view, hangars and offices, the support units that are needed to keep a military base functioning. My escort shows no sign of slowing down, or of arriving anywhere for that matter. The distance we are driving surprises me slightly but it probably shouldn't in the country that does everything BIGLY. Including offences against grammar.

A couple of days later an American airman explained why there was such a long drive from the main gate to the airfield. Local legend has it that when the Predator squadrons were first activated at Creech, the USAF personnel who operated them needed to live in Las Vegas. The distance from the edge of Las Vegas to the airfield was just far enough for a particular home-to-work mileage allowance to kick in, at considerable benefit to the commuters and considerable expense to the government. This prompted the new entrance to be built next to Highway 95, taking the distance from Vegas to Creech officially below the mileage allowance threshold. I decided not to try and verify the story because I did not want to discover that it was untrue. Even if it is an urban myth it somehow speaks to an ageless truth I have encountered in several countries: if there is a financial allowance, personal benefit or some other factor that makes life more tolerable for military personnel and their families, someone, somewhere, is working out how to remove it.

Eventually a sign indicates life: 'Home of the Hunters, 432nd Wing, United States Air Force.' Somewhere ahead lies 39 Squadron, RAF – a lodger unit amongst the permanent American residents.

Approaching the squadron area and taking in the view, my mind is prompted towards a scene from the film *Independence Day*. Desperate survivors of an alien-invasion apocalypse descend upon the secret government programme at Area 51, which, ironically, is not very far from here. Massive multi-layered security systems protect a secret world where giant, shiny glass and metal doors give way to a brightly lit, futuristic complex where all manner of other-worldly weapons and technologies are hidden. An army of white-clad scientists and military specialists work on projects beyond imagination and beyond accountability. Even the President is kept in the dark. My excitement level rises.

Then a seemingly invisible but shockingly effective speed hump jolts my head into the car roof and shocks my mind back into the present. As my escort guides us into a car park my anticipation sensors rapidly reconfigure from the excitedly overwhelmed to simply being 'whelmed', before heading rapidly for the distinctly underwhelmed. Why? While the sign on the chain link fence behind the car park says '39 Squadron, Royal Air Force', the buildings behind the fence do not yell out 'Area 51 Futuristic Facility' or even 'Secret Hi-Tec Drone Lair'. They just mutter 'Dusty Portacabins with some sand-coloured shipping containers lined up outside'. It looks like an RAF detachment in 1920s Iraq.

The car park itself is almost a caricature of the Brit abroad: a smattering of American muscle cars and motorbikes hint that several boyhood fantasies are being lived out by some nearby ground-dwelling *Top Gun* wannabees. Tom greets me cheerily with that most British of welcomes, 'Dr Lee, I presume!' Livingstone and Stanley in the searing Nevada heat.

Tom has, no doubt, been volunteered to make the practical arrangements for my arrival. He has done the job well and hands over a folder of useful information, the most important of which is a provisional list of people who have agreed to be interviewed. I heave a huge sigh of relief; I'd travelled all this way with only *one* confirmed interview in place. The second thing he hands over is a swipe card and security code number. 'This will get you in everywhere. Don't lose it.'

I played it cool and nodded knowingly: 'I'll be careful.' But inside I was thinking, 'Secret shiny drone lair *here I come*.'

The revolving security gate is particular effective at keeping out intruders. And at keeping out those who should be getting in. Multiple swipes of our passes are accompanied by beeps, red lights, muttered curses and a complete lack of access. We shall not enter. There is something mildly ironic about the being able to fly a Reaper thousands of miles away while being thwarted by a temperamental electronic lock.

Some helpful assistance gets us past the blockade and into the Squadron Operations building. The main door opens into the crew room and tea bar. This is definitely not a secret shiny drone lair. If the interior designer is aiming for jumble sale chic with shades of unwanted military supplies, all decorated with uneven standard issue framed pictures of the Reaper and other RAF aircraft, then the look is a triumph. The ambience is completed by the lingering scent of microwaved curry, the sound of the BBC News channel and an air conditioner losing its rear-guard battle against nature.

I quickly lose track of the introductions and names. The daily briefing is due to take place in a few minutes and nobody is loitering. More instructions as I am steered towards an internal security door. My electronic equipment – laptop, tablet, digital recorders and mobile phone – is abandoned with everyone else's outside the SECRET operations area. I grab my notebook and pen. The swipe card works first time and gets me into the secure zone. More introductions and the occasional promise of 'I'll speak to you later.'

Around me, last-minute preparations are taking place for the briefing. To my left the Duty Auth's (Authoriser's) desk would be familiar to anyone who has ever visited an RAF flying squadron. Any lingering vision of a chrome and glass wonderland is replaced by the reality of varnished plywood topped by Perspex: a classic air force design. Its angled top surface holds the documentation for the Reaper aircraft that is thousands of miles away, but which the pilot will later sign for before taking command of it.

In the Ops Room across the corridor from the desk, some footage of a strike the previous day by XIII Squadron in the UK is being examined. I ask a passer-by if there is something unusual about this particular video but it turns out to be routine: footage of every weapon release is shared between the squadrons for training purposes. Various levels of critique are being offered involving angles of attack and weapon settings, all underpinned by a general sense of '39 Squadron would have done it better.'

I smiled and thought: *Everyone's an expert.* My second thought was: *Actually, everyone watching is an expert, and they are analysing the minutiae of destroying and killing in forensic detail.* I can just see past the small group huddled around the screen but cannot quite fathom the details being discussed. In a few weeks' time an instructor at RAF Waddington would talk me through a broad selection of different weapon strikes, what the crew are looking at, how the pilot is flying the aircraft to set up the shot, and so on. I'm not sure how I would feel about every lecture I deliver being dissected, line by line, by all of the other lecturers in my department but I am certain I wouldn't like it.

The screen goes blank and everyone makes a final move for the seats in the Briefing Room. A junior officer is on the receiving end of some banter about the length of his hair. He shrugs it off with threats of revenge and vague promises to get to a barber. I am struck by the banality of the exchange. The comments seem simultaneously appropriate, given that this is a military unit, and inappropriate – given the significance of an extra 5mm of hair in the context of what the squadron will do today.

The room itself is nondescript and functional, maintaining the 'every expense spared' theme of the building. A few maps enliven the walls. The INT board with the latest intelligence updates reminds me of the local Items for Sale boards in many supermarkets. Useful for the right person at the right time with very specific needs, but otherwise generally disappointing.

The focal point is the lectern where last-minute adjustments are being made. Beside it, the screen onto which the first PowerPoint slide of the pre-flight briefing reminds us where we are. The mission crews are going to spend the day, mentally at least, in another country and in a different time zone. The relief crews – who will step in to provide rest and lunch breaks for the mission crews – could be operating in two different places throughout the ten to twelve hours ahead, depending on what and where the missions are.

When we start to take our seats the bonhomie subsides. Notebooks appear, quiet descends and attention turns to the screen. Everyone stands when the Commanding Officer enters. As he walks past he taps me on the shoulder and says quietly, 'You're flying with me today.'

'Great,' I whisper in knowing agreement. I don't know exactly what he means but things are getting more interesting by the minute.

'Good morning sirs, ma'ams, ladies and gentlemen.' The Auth, who must have at least 500 hours' experience on the Reaper to qualify for this particular duty, begins the briefing. It is one of several responsibilities he will hold for the duration of his duty period. The most important of these is being legally empowered by the Station Commander, via the Squadron Commander, to ensure that flying and legal standards are maintained throughout the day's missions. He will sign the crews out, sign them back in, be available for advice and, in between, monitor what they are doing through the live video feeds from their respective aircraft. He has hotlines to the Command Headquarters, lawyers and anyone else he might need to speak to.

The briefing replicates that found in every RAF squadron, for every type of aircraft, everywhere in the world. It begins with the MET (meteorology) briefing, which provides a detailed weather forecast for the transit route of the Reaper, from its launch site to the day's area of operations over IS-held territory near Sharqat in Northern Iraq. The next image on the screen depicts what the wind, cloud and temperatures will be in the area for the next twenty-four hours and at different flight levels. The daytime reading can be summed up as 'hot and sunny' followed by 'hotter and sunnier', all interspersed with limited viz (visibility). A thirty-year-old memory resurfaces. I recall drawing a similar MET map on acetate as a university air squadron flight cadet at RAF Leuchars in Scotland and presenting it on an overhead projector. It probably said 'cold and cloudy' followed by 'colder and cloudier,' all with zero viz.

Important information follows, signposted on the screen by 'IMPORTANT INFORMATION' for the inattentive. A general recap of the last few days of the campaign against IS follows, updating those who are returning to work after a couple of days off. Actually, instead of IS the Auth actually uses the term 'Daesh' – an Arabic term with vaguely insulting undertones – as mandated for all UK government departments. Recent activities by both 39 Squadron and XIII Squadron are summarised. Short videos are played of the destruction of both a pickup truck with a weapon on the back – mobile light artillery known as a 'technical' – and a bunker system. 'Squirters' appear from both strikes.

Squirters are enemy fighters who somehow survive a missile or bomb strike and run off; depending on the RoE they may be re-attacked by the crew. It is a lucky squirter who lives to tell the tale. I try to imagine how disorienting it must be near the blast of a 100lb Hellfire missile when it impacts at many hundreds of miles per hour. I have a sneaking admiration for the survivors who have the presence of mind to run and hide.

A quick summary of aircraft serviceability follows: no problems today. That will shortly turn out to be a bit optimistic.

The dedicated crews for the two 'lines' are identified. A line is shorthand for everything involved in a single mission from take-off to landing. One line will be devoted to direct support for

Iraqi forces fighting against IS. The other will support operations against IS in Northern Syria.

Flex crews are also detailed. Their job is to relieve the duty crews for their breaks and meals by taking temporary charge of the Reaper in question. They dutifully write down the planned timings. The German strategist Helmuth von Moltke is attributed with the words, 'no battle plan survives first contact with the enemy'. That extends to the Reaper crew meal plans.

Finally, the specific briefing for today's operation generates a noticeable increase in interest. The satellite image of the operating area is annotated with lines, arrows and various indicators of the enemy disposition, according to the most recent intelligence reports. I wonder why the Auth didn't just use the regional map I looked at yesterday on the BBC website; it definitely seemed much clearer. An arrow at the top of the slide points upwards with MOSUL typed next to it, giving me some rough bearings. It would be many months until the battle for Mosul got under way. Different phases of the operation are detailed in relation to Iraqi Army progress, or planned progress, on the ground.

All of this takes place at breakneck speed, much closer to that of a horse-racing commentator than to a bingo caller. A quick 'Any questions?' is followed by a few clarifications. Then, before I know it, the whole thing is over. My neighbour tells me that the briefing is kept shorter than that at XIII Squadron at RAF Waddington because 39 Squadron will be flying long transits to the operating area, and further information can be passed on to the crews then.

Everyone rises for the departure of the Squadron Commander, i.e. the Boss. The crew I am to shadow will be walking to the GCS in fifteen minutes. On the way out the pilot tells me to get a drink and a toilet break: it will be nearly three hours before the next opportunity. I wrestle with dehydration as a preferred option if I am three hours from the next toilet break. I definitely don't have the bladder to fly a Reaper.

I leave the Briefing Room somewhat dazed by the speed and amount of information given. One thought dominates. In this mind-bending world of remotely piloted aircraft, the war against IS is roughly seventy-five feet from where I am standing.

CHAPTER 2
WATCHING

'THIS IS AN AUSTIN POWERS PARKING MANOEUVRE RIGHT NOW.'
DEAN, REAPER PILOT

DAY 1

I exit the building through the crew room door with an empty bladder and a full bottle of water. By the time I return in a few hours the water will have swapped locations. I don't know why I am preoccupied with bodily functions. My mind is on overload, trying to take in everything I have seen and heard in less than one hour. Open cups or mugs are forbidden because accidental spillage in an electronically dense environment could cost huge sums of money in damages and vital lost hours of operational capability.

The desert heat hits me and I almost laugh at the extreme sensation. Such a temperature has never been experienced in my native Scotland. I am wearing a short-sleeved shirt and after a few seconds my pale forearms feel like they are being assaulted by the sun's rays.

It is only a few yards to the entrance to the first of four sand-coloured GCSs. They are lined up and spaced out with military precision, all under a canopy that tries to protect them from the direct heat of the sun. An air-conditioning unit hums gently next to the first container, where I will spend most of the next ten hours. Black arteries snake into the GCS carrying electricity, audio and video signals, telephone lines and millions of digital 1s and 0s per second that make up the complex, secret computer coding that makes it all work. For the most part.

Over the years I have seen many, many pilots and other aircrew walk out of their squadron buildings to their aircraft. From Norway in the Arctic Circle to Scotland, England, Gibraltar, Cyprus and the Falkland Islands the pattern is largely the same. Some similarities stand out as I watch the Reaper crew walk to the GCS. The Reaper guys are wearing standard flying suits with 39 Squadron patches and relevant rank slides. They carry themselves with the kind of confident nonchalance that has been the mark of aircrew for more than a century. The same confident nonchalance that bugs the crap out of non-aircrew the world over. Some differences stand out: they do not have flying helmets; nobody is wearing the G-suits that help fast jet crews resist the effects of gravity; and there is no waterproof layer. This Reaper crew will not be crash landing or ditching in the sea if there is an in-flight emergency.

But something about them is off, doesn't make sense. Something does not quite fit as the three of them file in to the metal container ahead of me. As I follow behind my subconscious dredges the answer from somewhere. Two of the crew are actually carrying cold weather flying jackets with them, in 100 degree desert heat that is quickly turning my pale blue shirt into a damp, dark blue dishcloth.

As I step inside, the temperature drops 40 degrees to around a steady 17°C (62°F). For me this is pleasant; for the acclimatised Reaper crew it feels like winter and a jacket is an essential requirement. The temperature drop is accompanied by a sudden gloom as I enter what I first imagine to be the inside of a giant computer from a 1980s sci-fi film. A wall of computer equipment faces me in the narrow corridor that runs about 15ft from the pilot and SO who are taking their seats to my right, to the MIC who is sitting in front of his own screens to my left at the back of the cabin.

The SO beckons me to a seat in between, and just behind, him and the pilot. I close the door, shutting out the Nevada desert. Another desert landscape will soon occupy our attention. In the meantime I am fascinated by the fact that the walls, floor and ceiling are all carpeted. Not high quality woollen Axminster carpet, more the kind of hardwearing industrial weave. I make a mental note to ask someone what the carpet is for. I make a second mental note that there are more important things going on.

I am handed a set of headphones with a chord that looks long enough to reach right back to the MIC station. As I adjust them for comfort I can hear that the pilot and SO are already running through their pre-flight checklist. I look up to see a bank of around a dozen small television-size screens in front of them, with four smaller screens. In between them are two old-fashioned telephones.

Then the pre-flight checks grind to a halt. There is a problem with one of the live information feeds. The Auth in the Operations Room next door will not approve the start of the mission until it is sorted out, which could take a couple of hours. There is so much information flowing into the GCS from different sources that I am intrigued that this one problem is a show-stopper. During a lull in the conversation the SO

– who also happens to be the Squadron Commander or Boss – explains the situation in terms he thinks I will understand. 'Basically, we have lost something like the equivalent of a car satnav. The alternative is to have a map plus verbal updates from elsewhere, as required.' He has seen this problem before and is confident it will be resolved quickly.

I also note how the personnel dynamics start to shift. When the Boss is in the GCS as an SO crew member, he is subject to the supervisory authority of the lower-ranked Auth in the Ops Room next door. In this instance, the Auth will not approve take-off as the lack of a 'satnav' could potentially reduce crew situational awareness. The Boss is confident that the problem will be sorted before the Reaper reaches its operating area in Iraq and that it is safe to send the aircraft there. However, it is the Station Commander back at RAF Waddington who legally 'owns' the risk involved and only he can allow the Boss's plan to proceed. It will have to be the Boss who phones the Station Commander, wakes him up in the middle of the night in the UK and asks him to give permission to proceed with the flight. So, the Auth gives the Boss approval to leave the GCS to take charge of the situation temporarily; the Boss then phones the UK and gets agreement for the flight to transit with the limitation; the Auth then approves take off; and the Boss takes his seat again as the SO under the supervision of the Auth. Simple.

It all seemed quite convoluted but the underlying principle is that the person with supervisory authority over the aircraft is responsible to the legal owner of the risk; in this case the Station Commander. If the Auth had simply decided to launch the Reaper without getting authorisation from above – and if anything then went wrong – it would have been his neck and career on the block.

Flight preparations proceed an hour behind schedule.

Everybody seems happy with the outcome except, perhaps, the Station Commander in the UK who is now probably trying to get back to sleep.

'So, what happened to "kick the tyres and light the fires"?' I ask. The old-school fighter pilot adage.

A crisp, 'Very funny!' from somewhere tells me to shut up.

The checks restart and they will mostly be familiar to anyone who has flown anything from a light aircraft to a jumbo jet: airframe checks; engine checks; area of operations and maps; comms (lots of different types in this case); clocks (important when working across several time zones); radio frequencies; and weather.

There are also numerous system checks that differ from conventional aircraft and indicate that this Reaper is to be piloted remotely. The 'command link' is one example. It connects the pilot's and SO's controls in Nevada to the aircraft via fibre optic cable and satellite.

Then there is the preparation to electronically take control of the aircraft. Somewhere at a runway within a reasonable transit time of the Reaper's operating area in Iraq, a separate Launch and Recovery Element (LRE) will ensure that it takes off and lands safely. It looks like a vastly bigger version of a radio-controlled model aircraft take-off. Once the Reaper has safely climbed a few thousand feet into the air, the flick of a switch diverts the electronic control signals from the LRE crew in the Middle-East to the crew members in front of me, who are now actively controlling the Reaper.

As the Reaper continues to climb I can work out most of what I see on the pilot's screen. To the left of his main screen the aircraft's indicated airspeed is around 110 knots. Ominously, the number 88 is highlighted in red. This is the stall speed at which there is not enough lift from the air under the wings and the forces of gravity take over. Bad things happen after that. There are also a couple of unfamiliar hieroglyphics, so I ask.

'These markings indicate the weapons I have available. Four Hellfire laser-guided missiles – two on the left and two on the right – and one GBU-12 guided bomb.' The former weigh 100lb each and the latter 500lb. Either can be immensely damaging or immensely helpful, depending on who you are and what is happening at the time.

Meanwhile, the SO is checking the camera pod in its 'normal' and 'infrared' modes. Later he will check that the laser for guiding the weapons onto their targets is working. For now, however, he catches me out: 'Do you want to jump in the seat and try the controls?'

'Sure.' As we swap seats I know that I am about to impress the hell out of him. I have flown light aircraft in the past and have also spent many hours on computer games.

The left-hand control allows me to zoom the picture in and out. He gets me to zoom in quite close to watch a car driving on a road that cuts diagonally across the screen in front of me. The right-hand control moves the crosshairs in the centre of the screen.

'Now keep the crosshairs on that car,' he tells me, pointing at the screen. 'It's perfectly safe,' he adds, reading my mind. I am sure I hear a chuckle from the pilot or MIC, but I am about to show them some moves. Some really bad moves, as it turns out – at least initially. The left and right movement is quite straightforward, but the up and down control seems to be the wrong way round. Worse, the joystick seems unusually big and clunky and not nearly as good as some on computer games. When I mention this later I am told that the manufacturers used old F-16 controls that were surplus to their requirements. And I believe it.

If I was laser-guiding a missile at this point, the safest place in the world would be in the car that was now dancing all over the screen. Everywhere except near the crosshairs. In my defence, the problem was made worse by the one to two second time delay between me moving the control in Creech and the satellite delivering the signal to the aircraft. It took a couple of minutes for me to start to coordinate the moving, time-delayed, three-dimensional challenge in front of me. It took roughly the same time for the others to stop laughing.

Before I know it my familiarisation exercise is over as the flex crew arrives to take over temporarily and give the duty crew a beak. The delayed start of the mission means we are not yet into the designated operating area. The Reaper will get there by the time we have eaten and return.

The Squadron Commander reverts back from SO to Boss mode. He goes to confirm with the Auth that the 'satnav' capability has returned as anticipated, and to find out what caused the problem.

A short walk and a couple of security barriers away is the chow hall, or what the British call the mess or canteen. Hot and cold food is available twenty-four hours a day. It is a barn of a building that can feed hundreds of people at a sitting, at tables that are laid out in precise rows. Americans are often criticised for their diet and I am impressed to see some crisp salad in the fridges alongside the cheesecake, opposite the counter where burgers, steak, pizza and other local delicacies are churned out. Just as I am about to rewrite my prejudices about Americans and food, a young airman comes over and grabs a handful of salad to add some colour and texture to his massive double burger and cheese. Or perhaps to help his stomach process the pound of beef that is coming its way. From his physique I do not have to be Sherlock Holmes to work out that this is his standard diet.

My two hosts from 39 Squadron regale me with tales about the merits of living and operating at Creech. It boils down to this: Creech is a base that is at war and everything is available twenty-four hours a day to support that effort. I will get to experience what happens at RAF Waddington in a few weeks' time.

Re-entering the GCS, each of the crew gets a quiet update from their counterpart in the flex crew as they settle back in for the next few hours.

'Eyes on!' The change in the pilot's tone is enough to tell me that the transit is over and we – or at least the Reaper that the crew is flying – are overhead today's Named Area of Interest. For my benefit he adds, 'This is a Daesh area.' The use of the official UK term – Daesh – does not escape my notice. It feels out of place, forced, like middle-aged parents using text-speak to – LOL – get down with the kids.

From the rear of the GCS, the MIC outlines the latest intelligence picture as he absorbs the information from multiple inputs, from text chatrooms to visuals and direct audio instruction. These include the observation priorities from the Combined Air Operations Center (CAOC) at Al Udeid Air Base in Qatar. The CAOC provides command and control for all air assets – the different kinds of aircraft – operating in Iraq, Syria and Afghanistan.[10]

There are two immediate priorities, though if emergencies arise those priorities can change at any time. The first is to check the site of yesterday's missile and bomb strike against IS to see if activity has resumed in the location. The second is to search for a couple of IS technicals that have been seen in the area. These mobile weapon-carriers can have anything from 0.50in-

10 Further details about the CAOC can be found at http://www.afcent.af.mil/About/Fact-Sheets/Display/Article/217803/combined-air-operations-center-caoc/, accessed 30 July 2017.

calibre machine guns whose rounds can penetrate concrete, to larger and more devastating anti-aircraft guns. I don't need to spell out their primary job but they also make very effective and fast-moving artillery. Hit, run and hide is the maxim. They conduct the rapid, aggressive manoeuvre warfare that IS used so effectively to seize as much ground as it did in the early stages of its offensive. That was before Western air power joined the fray, especially the Reaper with its exceptionally long loiter times, surveillance capability and weapons.

The screens before me show slow-moving images of bleak, sandy countryside, punctuated with random houses, settlements and towns. When we arrive at the scene of yesterday's strike, it seems clear to me that there is nothing going on. Maybe I just have a low boredom threshold but I would have been away from that area in two minutes flat. Rubble is rubble and the damaged building nearby looks uninhabitable. But then, perhaps, my idea of uninhabitable is different to those who are fighting a war for which they are ready to die.

The self-sacrificial element of what the IS fighters are doing is difficult to ignore and even more difficult to understand. Dying for one's cause is the ultimate commitment. There's also what they do to innocent civilians and to Muslims from different historical traditions. And this brings me to a curious corner of the public drone debate and a word that regularly crops up in discussions about drones: 'fair'. As in, 'Is it fair to use remotely piloted Reapers against jihadists who can't strike back at them?' Hilarious. The notion of war as a fair fight has emerged somewhere in recent arguments against the use of Reaper. Since the time of the Chinese military theorist Sun Tzu more than 2,000 years ago – and probably before – the idea has always been to make war as unfair on your enemy as possible. The advantages offered by RPA are not a violation of traditional military strategy – it is what militaries have been after for centuries. If activists want to use drones as a kind of lightning rod for anti-Western, anti-technology, anti-globalisation, anti-government criticism they should just say so. Instead, they can drown out the dedicated, informed scholars, activists and journalists who work to hold governments to account over their use of Reaper and other RPA.

Here's a test. Look at wars throughout history. Start listing the ones where political or battlefield leaders deliberately surrendered a distinct advantage to give their enemy a fair chance of winning. It will not take you long. (And no, I am not talking about the occasional act of chivalry or compassion on the battlefield.)

Anyway, back to the screens. Yesterday's reconnaissance of the area had been limited by strong winds and the sand it dispersed in the air. Today, in contrast, there are crystal clear skies and maximum visibility. The thorough examination of yesterday's strike site yields no indication of life and activity. Everyone agrees that the job is completed. The MIC receives a new tasking, or task, and gives the pilot a bearing and destination. But the pilot does a curious thing. The direction indicator on the screen shows that the Reaper has responded to his joystick. Specifically, I can tell that he has gone into a left turn. In an aircraft cockpit in flight, gravity and centrifugal forces combine to cause the crew to lean into the turn. Despite this particular cockpit being a shipping container that is firmly anchored to a concrete base in the middle of the Nevada desert, the pilot still leans left into the turn. I don't know if his brain is telling the rest of his body to move that way, or whether he is using his body as a means of understanding the movement of the distant aircraft. Either way, it looks weird.

Almost immediately the SO spots something and zooms in the camera on his pod. A technical fills the screen. It is a quad cab pick-up truck and the MIC identifies a 0.50in-calibre machine gun on the back. Nobody seems especially interested or excited about this development. The SO explains: 'This is not a Daesh-held area so it is unlikely that this will be one of their vehicles out on its own. We are just working out who it does belong to.'

The three crew members each contribute description and analysis of the image on the screen. Even I can tell that the markings do not belong to IS. The MIC breaks away from the conversation to check on the several intelligence chatrooms he has running on his screen. He types in a description of what they are seeing and where it is. The MIC is not just a recipient of intelligence, he is also contributing to the overall intelligence picture of the area. Some quick cross-checking confirms that, in this rapidly changing environment, the technical belongs to a non-IS, non-government militia.

'Does that make them the good guys?' I ask. The question is intentionally mischievous. There is a long silence. The pilot is the first to respond. 'Well, they are good-ish – at least for now.' His hesitation is understandable. 'Good' and 'bad' are much-misused, malleable terms in this conflict. As are the names and allegiances of many of the militias involved in it.

They move on from the good-ish, bad-ish technical to get on with the systematic area search south of Mosul. After half an hour of 'watching buildings go round' as the camera moves from building to building and small settlement to small settlement, I find my attention drifting slightly. There has been no sign of life in the last three places we looked.

'Where is everybody?' I ask.

'Were you watching BBC News last night?' came a reply. 'The luckiest ones have family in safe areas. The other lucky ones are in a refugee camp or trying to cross the Med. by boat to Europe.' The images of refugees being rescued from their pathetic, barely seaworthy boats are harrowing. There is something similarly shocking about the empty places that the refugees have left behind. Homes, schools, businesses, mosques. The spaces that were once filled with life and vibrancy are now empty shells, many destroyed or damaged in the fighting.

While the discussion of the empty homes was going on, the MIC was receiving updated information from one of his chatrooms. He gives the pilot and SO an eight-figure grid reference – still some miles south of Mosul but further to the west – which the CAOC would like to investigate. Intelligence from the myriad of supported forces on the ground in Iraq has made its labyrinthine way to Nevada for visual confirmation from the air.

However, before the crew reaches the location the flex crew reappears. Dinner time. The MIC updates everyone on the latest intelligence picture. Each crew member then gives a detailed description of what they are doing, what they are looking for and why. Then one by one the members of the relief crew settle into their seats. One hundred years on from the rapid growth of air forces in the First World War, control is handed over with the words that aircrew – often instructors and trainee pilots – have used ever since.

'You have control,' says the exiting Reaper pilot.

'I have control,' responds the relief pilot.

The crew head off separately to make phone calls, answer emails, you name it. For those with children, it might be the only chance to speak to them between school and bedtime.

One of the aspects of being a Reaper operator that is widely known is the disjuncture between home and work, particularly when work life involves observing often harrowing events and killing people who have been watched for long periods. What is not appreciated, and certainly did not cross my mind before I was confronted with it, is that the mental transition between war and peace does not happen at the beginning and end of every day. It happens at the beginning and end of every stint in the GCS during the course of single shift. That is a lot of mental readjustment on a daily basis.

I head off to conduct an interview, my first here. There is a false start when my interviewee, an off-duty pilot, realises that he has arranged for the interview to take place in the briefing room – a SECRET classified location – but my recording equipment is forbidden. So we find a quiet room. Moments like this away from the GCS remind me of the seriousness of what is going on and that this whole squadron is at war, not just the people who are doing the flying.

These first few interviews are crucial as I begin to engage with people who don't know me and live in a continual 'need to know' mode. These opening interviews are particularly valuable in the practical information I glean and for understanding how the Reaper and 39 Squadron work. At this stage the interviewees are understandably cautious about how much personal information and experience they can risk sharing. That's partly because of the politically and personally sensitive nature of what they do and how I might use it, and partly – as one pilot puts it over coffee – because of the Lee Rigby effect. Lee Rigby was the off-duty British soldier murdered in a London street in 2013 by two radical Islamists.

Importantly for me, however, another factor is working in my favour. A lot of the RAF Reaper personnel are fed up with how they see themselves and their work presented in different parts of the media. After the first couple of interviews a trickle turns into something of a flood. At this stage they are very keen to tell me what they do and how they do it. Some even make an early foray into telling me what they *think* about what they do. A crude, initial summary: professional and proud. It will be some time before some of them eventually open up to me about how they *feel* about what they do. As the events to follow will show, this is too complex for a crude summary.

Before I know it, I am called back to the GCS – dinner is over. I stand half way between the pilot and SO as they go through their handovers and retake their seats at the front of the box, as the GCS is known, and the MIC who is doing the same at his work station at the rear of the box. Do I detect a frisson of excitement? I have a random question. Why is the end of the box where the pilot sits seen as the 'front', while the MIC sits at the 'back'? In an information-based war, maybe the intelligence coordinators are the 'front' end. I make a note to see if I can start an argument about this in the crew room some time. (The answer turned out to be yes, and very easily.)

Once the flex crew left, I saw at the centre of the main screens a building with some kind of lean-to or temporary shelter in a small, narrow backyard that led into an alley. The MIC had located a technical as it entered this residential area and watched as it drove to this place a couple of minutes ago. It bore the hallmarks of IS and they had been in the process of identifying the weapon on the back when the technical reversed under the tarpaulin.

We watched as four armed men emerged from where the vehicle was concealed.

'Suspicious,' observes the Boss, but suspicion without evidence does not take them to the point of asking for a strike. Also, the presence of children in the alleyway – never mind who might be in the house – ensures a positive collateral damage estimate (CDE) and therefore no strike. It is time to watch.

Having had around just four hours' sleep in the previous thirty-six, and feeling the effects of jet lag, I worry that I might doze off. No way. The adrenaline kicks in. Potential strike locations are being sought by the MIC. Not easy in a built-up area with children and adults milling around. The Boss zooms out the camera view so that a wider surrounding area can be recce'd for potential strike locations. Then everything changes. Quickly.

The technical starts backing out of its hiding place, the camera zooms in and a sense of urgency permeates the GCS. A large gun on a tripod comes into view. Two different voices identify it as a 0.50in-calibre machine gun. The MIC starts to confirm the precise type of gun, as well as contacting his various intelligence sources to confirm 'ownership' of the vehicle.

The vehicle begins to track slowly through the back streets.

'The driver is looking for something or somewhere but probably does not know the area well.' The Boss interprets what we are seeing for my benefit.

Then it stops, reverses and moves backwards and forwards a couple of times as the driver tries to get closer to the adjacent building. He ends up further away than when he started.

'This is an Austin Powers parking manoeuvre right now,' says the pilot. It breaks the tension for a few seconds. I'm not sure if the others are laughing but I have to move my microphone away from my face. The people in that vehicle might only have minutes or hours to live and, frankly, I am fighting back laughter. Get a grip. I am the kid who laughed in school assembly and in church, then grew up to laugh in funerals (unfortunately, often when I was conducting them) and when I met Prince Charles. I am not a nervous or anxious person, I just like to laugh: and the darker the situation the funnier I find things. Firemen, doctors, nurses, undertakers and anyone else who deals with death and tragedy would get the humour. It is not trivialising what might be about to happen. It is a kind of pressure valve, a way to cope with things that nobody should ever have to cope with.

The armed passengers jump out and enter the building. New strike estimates are made and permission to strike – if a suitable kill zone can be found – is discussed. However, the current location still has too many unknowns for a strike to proceed.

The pilot keeps up multiple dialogues with the other two crew members, the JTAC and the Red Card Holder[11] (the RCH authorises a weapon strike) in the CAOC. The JTAC is the link between the RCH and the pilot, and is the one who will say 'cleared hot' to the pilot. The RCH suddenly seems to be juggling plans A, B and C, depending on what happens next. None of those plans will have a happy ending for the vehicle and armed passengers if they end up in an open space devoid of civilians.

Several minutes later the fighters emerge from the building more rapidly than they entered it and jump into the vehicle. It quickly becomes clear that the group is retracing the route back to where they just came from. They re-enter the narrow alley and the driver angles the vehicle to reverse into the narrow hiding place. He recreates the Austin Powers driving scene again with several attempts to park. After the fourth or fifth attempt – by which time he was getting the rear half of the vehicle into the hide – the driver stops, gets out and walks round to the other side of the vehicle. The other armed men leave him to it. The driver bends down and picks something up. A quick review of the video shows that he has knocked off the passenger-side door mirror.

'That might not be the worst thing to happen to your vehicle today,' mutters the pilot.

If this was on YouTube it would get millions of hits. As I try to gauge the reactions of the others (are the pilot and MIC being more restrained because the Boss is in the SO's seat?), another larger technical drives into view.

'Twin barrels. Anti-aircraft gun,' announces the MIC immediately. I can hear the increased interest in his voice. This second pickup is dwarfed by the huge gun on the back, a gun

11 In July 2018 the designation Red Card Holder was changed to Green Card Holder.

that will work just as effectively against ground forces as a piece of rapid-fire field artillery. It stops near the hide but nobody gets out. The MIC reckons he is waiting for instructions.

Sure enough, someone comes running out of the building next to the hide, speaks briefly to the driver and then disappears back where he came from. Technical 2 starts to move.

Decision time. Follow the new vehicle or wait and watch the hidden one? I struggle to follow the different conversations I can hear though my headphones. The discussions relate to available intelligence, the threat posed by the gun to both aircraft and friendly forces on the ground and the imminence of the threat posed by both technicals. I think everyone is missing the only salient point but it is not my place to join in: the driver of the second vehicle looks like he could actually get the gun into a position where it could do real damage. The driver of the hidden technical is actually a greater threat to his own side and might well destroy his own vehicle.

'Follow the mover.' Someone at the CAOC makes the decision. They will call in another air asset to observe the hide and wait for further movement.

It is the nature of the Reaper's observation pod that makes the decision so urgent. When the camera is fully zoomed in, say for a missile strike on a target or for capturing details on the ground with the greatest clarity, it is compared by the operators to 'looking through a straw'. There is an extremely narrow field of view. With the camera fully zoomed out to give the widest view, it would still soon lose sight of one of the two technicals. Plus, with the camera zoomed out, it would be easier for the crew to lose track of the second vehicle in this built-up area.

The pilot starts working with the CAOC and JTAC, getting the necessary conditional permissions and constraints for hitting the anti-aircraft gun with a Hellfire missile. Just starting the process gets my heart beating faster. Everyone I speak to in all of my subsequent interviews describes the same physical response, and more, to the build up to a strike.

No strike is approved yet – too many unknowns. However, the paperwork is in place and can be activated within seconds of the vehicle clearing the civilian area.

The vehicle was moving through the small roads of the town faster than any other passing car or truck. Perhaps the driver senses he is being watched. He will certainly know that he will become a target once he is spotted. I wonder if he is a hired hand or if he is a true-believing ideologue with martyrdom on his mind.

'Do you sometimes think the IS guys are setting out to die?' Things are temporarily quiet so I venture a question.

'Some, definitely yes. Others, I would say no. You see it in their behaviour. Mostly, though, you can't tell.' Apart from answering me the Boss is scanning the area for potential kill zones, depending on the route of the technical.

With the edge of town only a few hundred yards ahead – and therefore the open space a missile strike would need – radio calls get quicker and sharper. The conditional 9-line – the approval process that contains nine steps or pieces of information – only requires that there are no civilians in the blast zone. The pilot and the Boss discuss possible strike angles and missile settings. Apparently there is a choice. Who knew? I guess that is the point of military secrecy. They still don't spell out exactly why they go for a particular choice.

Without warning, the technical brakes hard and swings left into a large compound, squeezes past a tree and under some kind of shelter built onto the side of a house. On a Scottish farm I would expect it to be where a tractor is parked overnight. Within seconds, a tarpaulin is draped over the front of the shelter. If the crew had not watched the technical being driven in there they could never have spotted it. A map grid reference is noted and the crew settles in to watch this building for the remaining time of the shift. Where have the hours gone?

I push my wheeled seat back to the MIC's station. He is rewinding and replaying segments of the video from the drive through town. Screen shots are magnified and he is consulting with the Senior Mission Intelligence Coordinator (SMIC) sitting in the Ops Room next door.

'We are looking at these tubes here,' explains the MIC, pointing to some shapes on the back of the technical next to the anti-aircraft gun. From this view, the possible options include: mortar tubes, shoulder-held rocket launchers, something I can't decipher from my notebook or plumbing supplies. Apparently the last option was not a joke. A plumber's truck would make an ideal vehicle for a technical.

Through my headphones, I hear a series of checks being read out. When I return to the pilot's and SO's station they are preparing for a mid-air handover of the Reaper to XIII Squadron at RAF Waddington. Their shift is just starting and they will now watch the site where the technical is hidden, waiting to pounce if it ventures out to fight. I don't know why, but I feel a sense of anti-climax as we traipse out of the box. I mention it to the Boss.

'It's unfinished business,' he replies. 'We have seen what those guns can do [against the Free Syrian Army and others] and it's not a good feeling to go home knowing this one is still out there.' I make a note to explore this further.

'So what happens now?'

'Now we go home, sleep for a few hours and come back and do it all again tomorrow.' That was not strictly true. After a twelve-hour day he would do at least one or two more hours in his office. A day's flying is almost a break from his main job as Squadron Commander.

My mind wanders with fatigue as the crew goes through the in-brief with the Auth and the pilot signs the aircraft back in. The Auth runs through a series of checks and his last question sticks with me: 'Are you fit to drive home?' (They are required to declare if they are not fit to do so.) It is now the middle of the night and I am wrecked. There is a whole series of rooms in a building nearby with clean, available beds for those who cannot face the drive home. I suspect that regular self-deception goes on at this desk when it comes to tiredness and fitness to drive. I have a list of questions but they will have to wait.

I learn over the coming weeks and months that life on a Reaper squadron revolves around sleep management. It also revolves around trying to switch off, knowing that dangerous people with dangerous weapons have been spared bomb or missile strikes because of the risk to nearby civilians.

As I make my way back to my car for the drive back to my hotel room on the Vegas strip I doubt my own fitness to drive. And I have not gone through the extreme adrenaline surges that the crew experienced in the build-ups to the shots that they were never able to carry out. This most fascinating of days has left me with a profound sense of unfinished business. There is no finality, no fulfilment of a job well done. No explosions, no killing of the 'baddies'. The part of me that loves to see a task neatly finished would be driven mad doing this job.

When I get back on the highway, the darkness of the desert is oppressive in places. The headlights of other cars and trucks have a hypnotic effect. A distant glow on the horizon grows steadily until the bright city lights burst into view. It is an assault on my visual acuity after a day in a darkened box watching greyscale images of a distant desert land. I console myself with the possibility that it will all make more sense tomorrow…

CHAPTER 3
KILLING

'THREE... TWO... ONE... RIFLE.'
REAPER PILOT

DAY 2

Not even the sound of a stag party trying to recreate the film *The Hangover* in the two rooms beside mine could have kept me from sleep by the time my head hit the pillow sometime after 0200. When my 0630 wakeup call comes through I am instantly awake, refreshed and eager for my next encounter with the Reaper and the people who operate them. A 3,000-calorie breakfast – a statistic the casino breakfast bar seems inappropriately proud of – restores me almost back to my usual near-hyperactive state and adds a layer of hardening to my arteries.

Starting at just after 0800, I have several hours of interviews lined up at different homes around the Las Vegas suburbs before I am due back at 39 Squadron for another day's flying. The first two interviews are with Sian and Sheena, wives of serving crew members, while the third is a joint interview with Liam, a MIC, and his wife Melanie.

I obediently follow my satnav and trust it to take me to the first house. Rather than switch off the interminably insincere female voice, I am happy to accept every bit of guidance as I navigate a strange city on, for me, the wrong side of the road. There must surely be a market for a Billy Connolly-voiced satnav complete with expletives and suggested hand gestures. Some of the drivers around me seem to have it already as they welcome me to Las Vegas-style driving.

I arrive at the entry to a gated community that is better protected than almost any British military establishment I have visited, except for the Faslane nuclear submarine base. I press the intercom next to a house number I have been given and wait. A few seconds later a woman with an English accent – Sian – answers, presses a button and the imposing black metal gate smoothly opens in front of me.

I pull up outside a two-storey house that is impressive by typical RAF housing standards in the UK, but smaller than most of the houses surrounding it. I come to learn over the following days that it will have a floor area of no greater than 2,500 square feet – the limit set out by the embassy for house rental. Oh, and it can't have a swimming pool or hot tub. That rule applies even if a cheaper, smaller house with a pool would save the government money.

Two days later another Reaper pilot's wife confirmed this through gritted teeth, describing how she originally found a house whose cost was well under the size limit and under the monthly housing allowance limit set by the MoD and the embassy. 'Great value for money in the area,' she explained. But she couldn't have it. The house had a small swimming pool ('*Verboten* – bad for the image of the military abroad') and it exceeded the floor area that her husband was entitled to for his rank and seniority. 'First world problem,' she acknowledged. A third family – a non-commissioned officer this time – was almost stopped from getting their house because the landing at the top of the stairs was deemed large enough to be used as a sleeping area. Thankfully, someone gets paid to check this stuff.

Sian banishes her exuberant dog to a distant room as her husband Fraser mills around in the background: getting ready for work, dealing with the dog, answering emails and all the other minutiae of life that never stop for anyone. Sian and Fraser have opted to be interviewed separately. Theirs is a story that will resonate with quite a few couples in the armed forces. Girl (Sian's sister) meets interesting boy during RAF officer training. Girl decides to play matchmaker and introduces her sister Sian to Fraser – the interesting boy. Sian naively marries into the military while thinking it sounds romantic. Ten years down the line, Sian is older and wiser. Sian now waits at home while husband goes to work and, from time to time, kills people. Before dinner. Sometimes before breakfast.

Sian is open, candid and keen to take the positives out of life. She has supported Fraser through the years when he used to fly Tornado fast jets. She is not disappointed that he is currently flying the Reaper from a metal box on the ground rather than flying a jet up there, with the risks that entails. She tries not to seem overly keen on this arrangement because Fraser loved his fast jet days and she doesn't want to appear disloyal to his passion for flying, but I think she is a lot happier than she is letting on. And that's a common dynamic between couples on the Reaper squadrons.

'How have you seen him coming home from this kind of work with the Reaper?' I ask, venturing into more sensitive territory.

'I would say it is different… he's more tired quite a lot of the time. They have the commute as well, and they have to shift their body clock again on the days off.' Fatigue is a common theme that emerges, not in a self-pitying way, more in a pragmatic 'this-is-how-we-live-with-it' way.

'How much do you know about what he does?' I ask. She laughs. I don't yet know if that is because she doesn't know or if she does know but doesn't want to say.

'I knew all about the Tornado – you see them flying around and they look good. But with the Reaper you can't just go and see what they're doing because it's a lot more sensitive. I had to sit him down and ask him what it is he's doing. There is a lot more he can't tell me about his work compared to what he did before. Coming out here I've met a lot of different guys doing different jobs, and met different wives and got more sense of what their partners do.' Sian definitely falls into the I-want-to-know-more-about-what-he-does category. I will learn, though, that not all spouses or partners feel the same way.

The issue of what information can and cannot be shared with spouses and partners is part of the challenge facing Reaper personnel as they move between war and peace time every day. With few historical exceptions, when fast jet crews are living at home they are typically training rather than conducting combat operations, so talking about their work has fewer security implications.

By the end of the morning's interviews someone has spelled out one of the reasons why so many of the crew members and their spouses and partners have volunteered to speak to me. The partners often want greater insight into what the Reaper crew members are doing on a day-to-day basis, and the crew members want an insight into what their partners think about it. My consistent advice is: '*You could always try talking to one another a bit more.*' It does not take the whole morning to work out that the elephant in the room is the 'killing' part.

Wives, and more recently husbands, of soldiers, sailors and

airmen have always had to deal with the fact that the person they are married to could be killing other human beings when they are off at war. Historically, however, that killing happened elsewhere. Physically and metaphorically, that part of military life has usually been separated from home life. 'What did you do at work today?' is a risky question to ask – or answer – if the ending of another human life is likely to become dinner conversation.

By the time I jump back into my hire car to drive out to Creech for another day's flying, I have some insight into how weird it is to leave behind a perfectly normal family home to head out to a place where highly abnormal things happen on a daily basis. By the end of the week a particularly eloquent veteran pilot sums up the dynamics involved: 'It's a mind fuck.'

I escape the city almost before I notice. I turn off Ms Satnav so that I can think in silence. I have not yet had the chance to interview any husbands or partners of the women who fly the Reaper – that will happen soon enough – but the wives I spoke to this morning were thoughtful, feisty and good-humoured. They also struck me as being highly self-reliant – a necessary quality in military spouses – the majority of whom are still wives rather than husbands.

I don't quite feel melancholy, but I am more sombre than usual. It is eight years since I left the RAF and the past few hours have given me a fresh insight into the sacrifices my own wife made to enable me to serve in the armed forces. Women can now fight on the front line in almost every arm of the British military. However, expectations of many of the civilian women married to military men have not changed too much since the twentieth century, and possibly not too radically since the end of the nineteenth century: move round the country or world as required; live in a house that someone else selects; don't decorate your house to your own taste; raise your children as a single parent for extended periods; get used to regularly finding new schools, doctors, dentists, childminders and friends; and sacrifice or disrupt your own career if required. (Note to self: send Lorna, my wife, some flowers.)

Wives of military personnel are as interesting and diverse as any other group of people, and over the years I have noticed that they share a number of attributes. I don't know if these attributes are innate and somehow draws them to a military environment, or if circumstances force them to adapt. I suspect it is the latter, mainly because so many of the women married to Reaper crew members gleefully told me that when they were young they were determined not to marry military men: 'Too egocentric… too much separation… too much moving around with the families… no chance to develop a career of my own,' are the main themes. And yet… and yet, somehow it works for most. Love is not only blind but stupid. (Note to self: tell daughters to keep away from military men.)

Roughly half-way from Vegas to Creech my thoughts start to drift towards what the next ten or twelve hours will hold. I wonder what happened to the two technicals that I watched for hours yesterday. Will I see them again today? Or has someone destroyed the vehicles and their guns, along with the people using them?

I pull up at the checkpoint at Creech Air Force Base but my entry is a lot less stressful on the second day. I have the correct temporary vehicle pass and a personal pass now stamped 'Unescorted'. I momentarily wonder what will happen if I deviate from my route to the 39 Squadron compound but a lack of time and a fear of getting shot keeps me on the right road.

In the morning briefing, today's Auth answers my questions about what happened to the two technicals from yesterday: no, and yes. No, I will not be watching them both again today because, yes, one has been destroyed. The larger one was 'prosecuted' by 'allied fast air' [a fast jet, not a hand drier]. A jet from one of the coalition nations was used to destroy the technical armed with the anti-aircraft gun. US aircraft outnumber the rest of the coalition's fleet put together so, statistically, it most likely that the Americans carried out the attack, even if nobody says so.

The other technical – the one driven by the jihadist Austin Powers – is still being watched. However, the crew that I am allocated to will not be the ones waiting for it to move. Word has probably reached Austin about what happened to his colleagues just a couple of miles away. His choice will be to keep hiding for a couple of days in the hope that nobody knows where he is. Or he can embrace martyrdom and go on the offensive. Either way, he'll be very lucky to be alive by the end of the week.

My crew are tasked with following up on some active intelligence about IS fighters near Sharqat in the Al-Shirkat district of Iraq. There is a strong, well-organised group of jihadists who are putting up heavy resistance to Iraqi forces on the ground. They are using classic insurgency hit-and-run tactics: short, aggressive and effective attacks followed by a rapid retreat, regroup and re-attack.

These tactics have proven to be impressively successful over the past two years as IS seized more and more territory in Iraq and Syria. But, in the open desert terrain outside the towns and cities, the jihadists are highly exposed to the dangers posed by the air power above them, from reconnaissance aircraft to ground attack jets.

Rather than follow my crew straight out to the GCS to watch them flying their allocated Reaper to its area of operations, I get the offer of a couple of interviews with a pilot and a MIC who have an office day and are not flying. I jump at the chance. I am still not sure how many people will talk to me – ideally I would like fifteen or twenty from the two squadrons – so I have to grab every opportunity. If I knew then that my final total would be nearer a hundred, I could have relaxed a bit.

By the time I go out to the GCS with my allocated crew at the end of their first break, I already have a greater insight into what the pilot and MIC are doing and why they do things in particular ways. I also learn that every position in the crew is the most important – if you happen to be in that position. Hundred-year-old banter between the different crew members is alive and well.

Pilots are the 'two-winged master race' – so named because they must be commissioned officers and they wear a two-winged badge on their chest. They, especially former fast jet pilots, are seen to commonly project a hyper-sense of self-assurance, which others occasionally interpret as arrogance. The number of times I have heard a pilot called an 'arrogant bastard'. And I hear it again in the coming days. The MICs seem to be more sanguine about pilots and their general worth. From the back of the GCS where the MIC sits, pilots are seen as 'stick monkeys' whose job is to take the aircraft to where the MIC tells them. The SO is the camera operator who *looks* at what the MIC tells them, and guides missiles or bombs onto the targets confirmed by the MIC. Meanwhile, for the pilots and SOs, the MIC is a strange, friendless creature who lives in the back of their box and who would otherwise be homeless and hungry with no real purpose in life. And those are the kinds of things they say when they *like each other*.

The incoming pilot is now briefed by the flex crew MIC during the handover. Two targets have been positively identified as enemy fighters and a 9-line has already been raised by the JTAC in the Command Centre. As part of the 9-line process, legal permission from the RCH will be confirmed. The two targets arrived by motorbike near a long-abandoned home that now serves as an IS meeting place or command base. However, the decision to strike is temporarily on hold. The pilot is told: 'They arrived from the east and their choice is to keep heading west or double back to where they came from. Since they are moving away from an ambush they carried out, we reckon they will most likely keep heading west on the same road.'

The idle motorbike sits by the side of the road. The SO checks his infrared heat sensor and the bike's engine can be seen glowing hot near the centre of the screen.

'Confirm 9-line,' the pilot says to the JTAC. Since the 9-line includes approval to use lethal force, the pilot needs to check for himself that it is in place. Something might have changed since his temporary replacement received permission.

The 9-line is contingent on the blast from any missile strike not hitting the building as well. 'We can't risk there being unknown civilians inside,' adds the SO, pointing to the screen and taking a moment to explain their thinking to me. 'We might need to wait until they have moved off.' Until we kill them. Those last words remained unnecessary and unsaid.

It is surreal to be discussing when and how the two men I can see clearly on the screens will have their lives ended. I become aware that my heart is pounding. I have conducted many funerals and I have seen many dead bodies. I feel a grabbing sensation somewhere in my lower abdomen. I would say it is a pain in the pit of my stomach but this ache, I am sure, goes beyond a mere physical phenomenon.

My feelings of dread are interrupted by a blast of light and heat from the opening door behind me and the arrival of someone else in the GCS. The pilot from the flex crew has returned.

'Safety observer,' he whispers as he squeezes past in the limited space and crouches right in front of me, between the pilot and SO.

The safety observer is – as far as I can glean – unique to the RAF Reaper Force, being an experienced instructor, pilot or SO. He or she comes into the box in the build-up to a weapon being fired to act as an extra set of eyes and – hopefully – a calm mind. Their job is to spot any potential error or oversight that can be made when the crew members are intensely focused on some aspect of a strike, with adrenaline and a fear of messing up coursing through them.

Today's safety observer is fully up to speed. He was in the pilot's seat until a little while ago and, since then, he has been watching and listening to events unfold on the live video feed in the Ops Room. He seems physically relaxed in a way that the crew members, and myself, are not. He asks a couple of questions and confirms that the 9-line is up-to-date. I learn afterwards that he is one of the most highly regarded safety observers. He lets the crew do their job, does some quiet checking and does not upset the dynamic. I will also come to learn that some safety observers are not averse to a bit of back seat driving.

'Missile 3.' The pilot anticipates that the safety observer wants to check which of the available missiles is being fired. He also confirms the missile setting, which can be adjusted according to what angle it is being fired at and the kind of target that is being hit.

As the Reaper is being manoeuvred around the target I look for the pilot's turn-and-slip instrument to tell me how hard the aircraft is banking. Rather than a traditional dial, there is a digital display. It does the job but it is an odd sight – a thin digital symbol with all the finesse of a Commodore 64 computer readout from decades ago.

As I gaze at the pilot's screen, the three crew members –

with the odd input from the safety observer – discuss how best to hit the two men and whether they can do so while leaving the building intact. The easiest shot would be to hit the men at an angle where the building is behind them and in line with the missile trajectory. But it is also the shot where the follow-through blast could take out the front wall. Ruled out. 'Side' shots that keep a ninety-degree angle between the men and the building would also be straightforward, but the building is right on the borderline of the blast radius. Also ruled out. The most difficult shot would bring the missile in over the roof of the building and hit the two men. The laws of physics would send the blast away from the building but, if the SO guided the missile in just a fraction too low, he would remove a chunk of the building with a direct hit. Also ruled out.

'Target mobile,' the SO to my front right deadpans. One of the IS jihadists sets off on the motorbike.

'Roger.'

'Roger.' The pilot to my front left and the MIC sitting 10ft behind me confirm this latest development. The casual conversational tone of thirty seconds previously has disappeared. Imperceptibly, the crew's mood has changed, hardened.

I am sitting only an arm's length away, looking over the shoulders of the two men who control the Reaper thousands of miles away. In the semi-darkness my gaze shifts back and forth between the moving target at the centre of their screens and the two men who are watching it. The SO sits rigidly, eyes and hands focused on keeping the screen crosshairs on the moving target. The pilot rapidly repeats the well-rehearsed steps of weapon selection and confirmation of the missile settings.

'9-line update confirmed,' monotones the pilot. No trace of emotion. The crew have their authorisation to strike the target with a Hellfire missile.

I have never seen anyone killed before.

From over my shoulder the MIC is reading the roads ahead of the bike, calling out distances and times to the clean kill zone. He is also identifying the 'shift' area, the empty piece of land the missile will be guided onto if anyone unexpectedly enters the kill zone in the seconds prior to impact.

'Crew, all confirm happy,' intones the pilot.

Quiet checks and double-checks are acknowledged by the three crew members. I feel queasy. What are they feeling?

'Cleared hot.' The voice of the JTAC fills my headphones.

'Roger.' The pilot prepares to attack.

A countdown begins. 'Three… two… one. Rifle.' 'Rifle' indicates that a missile is being fired; 'stores' would have been used had a bomb been dropped.

The pilot gently squeezes the trigger. In the time it takes to read this sentence the Hellfire missile will accelerate off its rails towards the speed of sound.

For a moment, silence. I become aware that I have stopped breathing. Through my headset, I become aware that everyone has stopped breathing.

The cross-hairs are fixed on the image of the moving bike.

Still silence.

I look at the SO's hand gripping the joystick. His grip is so fierce I can see the whiteness of the skin round the top of his hand as blood is forced away from the area. The skin tightness betrays the calm voice. In my peripheral vision I see the safety observer clenching a fist. Part of his brain thinks he is flying the aircraft, or perhaps wants to be.

'Fifteen seconds.' That is how long the man on the bike has to live. I feel sick.

'Ten.' I stare at the crosshairs.

'Still clear,' adds the MIC. While I am fixated on the target he is still looking ahead, checking for civilians or unexpected events.

I brace myself for the impact.

'Five... four... three... two... one... Splash.' 'Splash' announces the impact.

The screen erupts in a white cloud as the Hellfire missile explodes. Then... nothing. There is no blast, no noise, no physical sensation to accompany the impact.

The instantaneous silence in the GCS is broken by a burst of exhalation from the SO. It had been a very difficult shot and, consciously or not, he had held his breath all the way through the missile flight until its impact. 'MY HEART'S BEATING OUT OF MY CHEST,' he exclaims, sucking in breath. His relief at hitting the moving target seems almost overwhelming. And before anyone else says anything, the pilot goes into a series of checks. It seems to take forever for the dust cloud to abate and for the aftermath of the strike to come into focus. Is it a minute? Two minutes? I have lost all sense of time and distance.

After gathering himself for a few seconds, the SO goes from missile mode to watching. His shoulders have relaxed and dropped a clear two inches. The colour returns to his hand as he eases the tension on the joystick. The MIC asks him to zoom the camera out for a quick check of the surrounding area before zooming back in to assess the results of the blast. Everything that is seen on the screen, said over the intercom or radio and written in the text chat boxes is being recorded for the post-strike debriefing that will come later.

As the cloud dissipates, the mangled bike is the first thing I can make out. Subsequent analysis will identify exactly where the missile detonates – front, middle or rear – but the crew are working it out. Then a body. Using the length of the bike as a measurement even I can estimate that the man has been thrown 10–15ft from the point of impact.

'Driver to the top of the screen,' observes the MIC.

While the camera focuses on the rider, the extent of damage to his body becomes apparent and the rifle is identified nearby. At first it seems that there are two rifles near the body but the MIC clarifies what we are seeing.

'It looks like his right arm has been severed and is lying next to the rifle.'

Oh God, it has. His arm must be a further 6ft away from the rest of him. Several post-strike activities continue almost simultaneously. As the scene is analysed from the screens in front of us, the pilot is speaking to the JTAC at the Command Centre. He is being put on standby for the next tasking. This Reaper will continue watching the blast zone – and for blast zone, read 'body' – but may be required elsewhere. I am told that colleagues of the dead jihadist will probably come at some point and take his body away.

As this discussion continues I become aware of a strong, putrid smell – hints of burnt flesh combined with surgical disinfectant. It doesn't make sense. We are in a climate-controlled, air-conditioned environment. I quickly look round at everyone. No reaction from any of them. But. But. I know this smell.

I cannot dwell on the odd pungent odour that nobody else is acknowledging. A shudder goes up my spine as I tune in to a change in events. The JTAC gives the crew a new grid reference just a few minutes' flying time away, where friendly forces have just been ambushed in a hit-and-run attack by a group of IS fighters.

The pilot sets a new heading as the JTAC starts a new 9-line authorisation. He will need to be ready to fire very rapidly if he needs to use a missile to disrupt an ongoing ambush. He writes key details in hieroglyphics in red on the whiteboard on the wall to the left of his head. He should be a doctor.

I move back to see what the MIC is doing. The key intelligence information is being funnelled through him and he is building a picture of the situation they are heading towards. Most important is the identification of 'friend or foe'. In other words, making sure this crew does not put a bomb or missile on its allies rather than the enemy.

He nods his acknowledgement of my presence and interest but otherwise ignores me. He is having multiple, concurrent typed chat conversations with his intelligence sources, and making sure his two screens are set up as he wants them.

'Zoom in, top left of screen.' With this experienced crew he knows the SO is listening out for his instructions. He breaks one hand away from the keyboard and touches the screen to show me what he is looking at. I note the traces of ink on his forefinger. As his hand returns to the keyboard it brushes the third of three pens slightly off to the side. He brings it into perfect alignment and spacing with the two pens next to it. A glimpse of order amidst the controlled chaos of activity.

To me, at best, there might be a couple of flea-size shadows moving ever so slightly next to a main road. There are so many other shadows being thrown by the rocky terrain I am sure I could convince myself of anything right now. As the camera focuses in on the requested area, the field of view becomes narrower and more detailed.

'Counting five so far,' he announces. He pauses the image on his second screen and rewinds the full motion video (FMV) by several seconds to double check the number of people he can see.

I roll my seat forward once more towards the pilot and SO. The safety observer has remained and is still crouching between them. He is waiting to see if they are going straight to weapons. I have not been told to move, but the MIC is performing mental gymnastics while being quizzed from at least three sources, as well as positively identifying the people on the screen as friendlies, enemies, non-combatants or some kind of combination. People might die based on what he confirms, and he does not need me as a distraction.

The SO partly turns to me: 'We'll soon confirm if these guys have been involved in a firefight.' He switches the camera sensing mode and the images on the screen transform instantly from a topographical view to a heat map. Rocks that have been soaking the sun's rays for hours stand out from those in the shadows. The fighters are no more easy or difficult to see in infrared, but they now look like they are carrying Star Wars-style lightsabres. Their rifles have all been fired very recently, recently enough for the barrels still to be glowing from the heat.

'If we need to hit these guys, we could use a 500 pounder,' says the pilot. 'Though with the narrow spread [of the potential targets] and the rocks around them, a Hellfire would do it.'

As they discuss possible weapon choices, the infrared exposes one possible further fighter in the outcrop.

'Zoom out for a wider view.' The MIC is looking for something – probably checking some intel or other. 'Can you move the camera as far to north-east as possible while keeping these guys in sight?'

'Roger.' The SO eases the camera pod over. They can't risk losing sight of them. The MIC quickly finds what he is looking for – the aftermath of an ambush about half a mile away with at least three Iraqi vehicles, or American vehicles used by the Iraqi army.

I try to interpret the discussion. The safety observer is watching and keeping quiet so I whisper, 'Is there another ISR asset involved?' (ISR refers to intelligence gathering, surveillance and reconnaissance aircraft like the Sentinel or Rivet Joint that gather up all kinds of radar images and electronic signals respectively.)

He nods his confirmation. I am none the wiser about whether it is British, American or someone else's ISR asset that put them onto the targets.

Between what he can see and what the other ISR aircraft can see and hear, the MIC gets the information he needs to positively ID the individuals we first spotted as enemy fighters. They carried out the ambush.

'Confirm 9-line,' says the pilot. No time wasted. Approvals take only a matter of seconds. The RCH has just been waiting for the positive ID on the targets to approve a strike. In the torrent of words that follow, 'Hellfire' stands out. They have decided to go with the Hellfire laser-guided missile.

The wider camera view also allows the MIC to carry out his CDE. It is easy on this occasion with nobody else around and nothing moving in the vicinity. The field of view shows desert and rocks with an empty road nearby.

It becomes impossible to keep track of the times 'check', 'confirm' and 'happy' are scattergunned in my headset. The 'business' tone once more takes over from the more relaxed manner of a few minutes ago. I realise with a start that I have not even thought about the first strike – the first time I have seen someone killed – for more than twenty minutes. Six more people are about to be added to that list. The queasiness returns.

I tune out the conversation about angles of attack and maximising the blast impact. A whole series of questions goes through my head. What are the fighters talking about right now? Are they scared? Exhilarated after killing some Iraqi soldiers? Wondering where their ride home has got to? Looking forward to dinner? Is anyone disappointed that he has not martyred himself yet?

The words 'cleared hot' from the JTAC to the pilot bring me back to what's happening.

Once more the pilot counts down, 'Three… two… one. Rifle.' He pulls the trigger and the missile is in the air. A few seconds of heavy silence dominates the headphones.

Whatever is gripping my gut is getting tighter.

I check the SO's joystick hand. The tautness of the previous moving target strike is gone. He just needs to keep the crosshairs static in the middle of the group of men.

I wonder why a joystick is called a joystick. It is wrong.

Concentrate. My mind is very reluctant to focus on the detail of what is happening. Does that make me a good person because I'm struggling to engage with the process of killing? Or a bad person because I am actually able to switch off from it? Or do I just have a short attention span? I certainly don't have any qualms that a group of IS jihadists are about die. They made their choices.

Then 'Three… two… one… Splash.' Silence. No applause, no cheering. At this moment I am very proud to be British. Not because the RAF has just killed some distant and deserving enemy. No, it is the complete lack of histrionics in the GCS. No whooping, celebrating or high-fiving. Just the best of British: self-controlled and emotionally repressed. There is a noticeable exhalation from the SO, but it is nothing like as percussive as his reaction to the more difficult shot against the bike.

The blast cloud fills the screen with greater intensity than the first missile strike, perhaps because of the surrounding rocks. As the camera is zoomed out for a wider view, the dust cloud keeps expanding. It definitely takes minutes to dissipate this time.

'BDA,' says the MIC, as calmly as reading out a shopping list. Battle damage assessment is top priority.

A crater appears at the centre of the screen. Bodies and parts of bodies are strewn around. I feel sick. I wallow in my patheticness for a couple of seconds as they start counting the bodies and describing a couple of large body parts that they can actually identify.

The putrid smell from earlier once again assaults my nose: burnt flesh mixed with surgical disinfectant.

As the safety observer moves past me to the door – his work is done for now – I whisper, 'Can you smell anything?' Even as I ask the question I think I know the answer. And why.

He shakes his head and shrugs his shoulders. 'Nope.'

I don't understand why I am smelling what I am smelling, but I realise exactly what it is and when I first smelled it. I encountered that unique mix of the smell of burnt skin and surgical chemicals in April 2003, soon after the start of the Iraq War. I was an RAF chaplain at the British military hospital in Cyprus, where British battlefield casualties were airlifted from Iraq.

My first, but not last, encounter with that distinctive smell occurred at the bedside of a young British soldier. An ammunition malfunction with a tank shell blew off his arm from the elbow down. I sat at his bedside trying and failing to find any words of comfort for him. Surgical disinfectant stained his skin just above the crisp white dressing that covered the site of the amputation. Perhaps it would be more accurate to say that it was the site where the surgeon tidied up the amputation that the blast made.

Some part of my brain is linking the hospital amputation from thirteen years ago with the images I am watching on the Reaper video feed, and dragging my memory of that smell into my consciousness. What is confusing is that *I am actually smelling something*. I would bet my life on it.

If this is what my brain is doing on my second day in a GCS, do other people encounter similar things? Am I just pathetic? Possibly. Am I just imagining it? Yes, obviously – there is no smell that the others can detect. Is the smell even real if it only exists in a series of connections in my parietal lobe?

I could drive myself mad with that question and resolve to find out more. I would later discuss this event with an academic colleague who specialises in memory function. He would very excitedly tell me about olfactory memory – recalling odours. Most people are aware that smells can sometimes trigger strong memories: of a dentist's waiting room; mulled wine suggesting Christmas; a sunscreen associated with a particular holiday. It turns out that powerful events can also trigger the reverse, which it did in this case. My brain connected the visual image from the screen with a smell embedded in my memory. It turns out that I am neither mad nor special. I just experienced a common phenomenon in a highly unusual place.

The following hours disappear in a blur of activity. The crew does not fire any more weapons today but they keep busy providing information for other manned aircraft that do. I think the explosion of a Hellfire missile is fearsome until I see a 2000lb bomb, launched from an unseen conventional bomber, explode.

By the time it gets to within an hour of the end of this flying shift I have a decision to make. I am mentally and physically exhausted after an eighteen-hour day, and emotionally wrung out with everything that I have seen. I am also still jet-lagged. I have to decide whether to drive back to my room in Vegas right now, which I think I can almost do safely; stay for the final hour and take my chances on the road; or stay for the final hour and sleep in one of the spare rooms. I opt to drive back right away as self-preservation kicks in. I say my goodbyes to the crew.

'Very sensible… Good idea.' Their words reassure me that I am making a wise choice. As I hit the Nevada fresh air I wonder if they are just being polite while really meaning: 'Pathetic lightweight! Can't stand the pace.' Or maybe I am projecting my own self-judgement onto them. I definitely need to sleep.

There is one final act in this most dramatic of days but my early departure postpones it until my return in the morning. As I walk into the crew room, three people ambush me and insist – quite strongly – that I sit down. I skipped town last night before the 'in-brief' at the end of the flight. I missed one of the mandatory questions, which they put to me now: Did I see any trauma risk management (TRiM) events?' That is, events which could be mentally disturbing or unsettling. If the answer is yes then the option of psychological support kicks in.

'Yes, I would say so.'

'Have you had any after-effects?'

Such as smelling things that are not there?

That's what I don't say. What I do say is, 'No, I don't think so.' Apparently any immediate responses can take up to seventy-two hours to occur. For a moment I wonder if I have just lied and decide that I haven't. The smell happened instantaneously, then went away – it did not appear afterwards and it wasn't distressing, just curious.

I promise to ring the squadron if I have any reaction, and they let me go. I note how easy it is to say what is necessary to avoid an awkward situation. I wonder if any crew members ever do the same.

This world gets more fascinating by the hour. As I hear and record more and more of the stories and experiences that span a decade of Reaper operations, I realise that I have just started to scratch the surface…

CHAPTER 4
PIONEERS

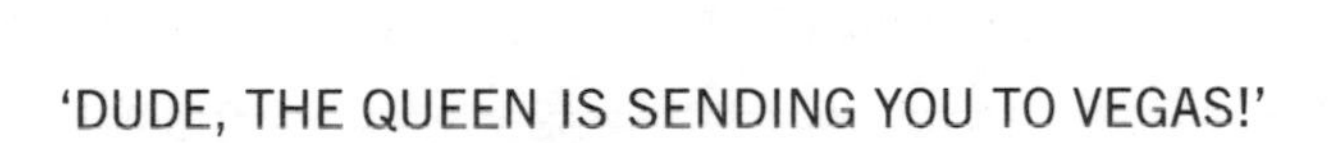

'DUDE, THE QUEEN IS SENDING YOU TO VEGAS!'

JOHNNY, PILOT

How do you know something exists? Deep philosophical questions are not normally associated with aircrew and flying squadrons, at least not until after Happy Hour starts and the scholastic benefits begin to be felt.

Future historians will argue about exactly how and where and when the UK's Reaper programme officially came into existence and No. 39 Squadron (RAF) was re-formed. These are not exactly the same thing. For example, the official RAF date for the re-formation of 39 Squadron is 1 January 2007.[12] But is a squadron *really* a squadron if it doesn't have an aircraft or a building? It took until October 2007 for flight operations to start, more of which later. Perhaps the formal 'stand-up' ceremony and the presentation of a *new* Squadron Standard – the ceremonial silk

12 See https://www.raf.mod.uk/organisation/39squadron.cfm, accessed 12 December 2017.

flag emblazoned with battle honours that is formally presented to a squadron by, or on behalf of, the sovereign[13] – marked the moment when a tangible symbol was formally recognised, and that occurred on 23 January 2008.

Alternatively, it could be all about the moment when the Reaper synthetic aperture radar (SAR) started to be used, in which case 2 November 2007 is the key date. The SAR became a crucial part of the basic targeting sensor equipment that provides the images for the Reaper to be used, effectively guiding the other high-tech cameras onto the target.[14] For some, the use of weapons is the factor that determines the operational effectiveness of the Reaper. In which case, there are two significant dates: 29 May 2008, when the first 500lb GBU-12 guided bomb was used to destroy a Taliban 81mm mortar position; or 2 June 2008, when the first Hellfire missile was fired against enemy fighters in an offensive named Operation GORDIE STRIKES.

For the purposes of this book, however, all of these official developments and significant dates merely provide a backdrop to the focus of action and fascination – the people involved. I am therefore going to propose that the *real* start of 39 Squadron happened on a hot, sunny day in 2006 when two Brits – both from the RAF – risked sunburn in the Nevada desert to plant a flag where the squadron building would soon be built. But even if I take that moment as a starting point, it was not so much a static, standing start to the UK's Reaper Force. Rather, it was one of many points on a rolling start that began a few years earlier

13 Squadron standards were introduced by HM King George VI on 1 April 1943, the twenty-fifth anniversary of the RAF's formation. The basic requirement for a squadron to receive a standard is completion of twenty-five years of service, but may also be granted to a squadron which 'earned the Sovereign's appreciation for especially outstanding operations' – a criterion met, for example, by 617 Squadron, better known as the 'Dambusters'.

14 For further detail see https://www.raf.mod.uk/rafwaddington/aboutus/39squadron.cfm, accessed 12 December 2017.

when the first British remote aircrew went to Nevada to be part of the USAF MQ-1 Predator programme, the predecessor to the Reaper programme. The experiences of several of those early pioneers give us an insight into how much air power and the RAF was changing, and perhaps how much it was staying the same.

PREDATOR

A classic RAF party pilot, Johnny was proud – in a way that some pilots are – of a career based on the self-deprecating motto of 'better lucky than good', i.e. safe enough and good enough to fly perfectly competently and consistently, but not quite good enough to compete for major career advancement, and always ready to try the next new thing. Johnny's career started in the early 1990s, just when the RAF experienced major cuts in personnel and equipment – the 'Peace Dividend', as it was known at the end of the Cold War. Like everyone else selected for pilot training, he first had to become an RAF officer. Then, while his peers spent a year or more hanging around in the UK waiting to learn to fly, he managed to bypass the entire RAF flying training system by getting an 'exchange' place in the USAF flight training programme.

Johnny spent a year in Texas flying T-37 and T-38 training aircraft, trying his best not to be a hazard to other users of the airspace. Having passed the course, he returned to complete fast jet flying training at RAF Valley in the UK. With a combination of a degree of talent, patient instructors, good fortune, struggling through the difficult sorties and repeating a record number, and being in the right place at the right time, he came through and, in 1997, was flying Tornado GR1 ground attack fast jets.

Johnny was soon combat ready, followed by an immediate

deployment to Turkey where he would fly over Northern Iraq and help to enforce the No-Fly Zone against Saddam Hussein's regime. At least he would have gone immediately to Turkey if he hadn't been bitten by Eric, the squadron's pet Rock Python. More interesting stuff happened to him over the next few years, including his involvement in the 1999 Kosovo air campaign. But anyone daft enough to be bitten by a Rock Python (twice) doesn't deserve to have all the details told.

During Operation TELIC – the 2003 Iraq War – Johnny was flying Tornados out of the main airbase in Kuwait and visited the USAF Predator unit. 'I'd gone in and seen the GCS and thought, *Oh my God, this is amazing, look at all the spangly buttons and stuff*. There was some pissed off US C-130 transport pilot who was sat there going, "Oh, I hate my life, I hate my job. I want to go home", but I thought it looked awesome – the future.'

For Johnny, the seed was planted. The Predator was now on his radar. Some months later, a friend of his called up: 'Look, I've been offered this job flying Predators in America. You've been to America to fly and I want to know what you think.'

'Dude, the Queen is sending you to Vegas, *paying* you to go to Vegas. Why are we even having this conversation? With the defence cuts, we're all going to be out of a job. Just go. And if you don't, I'm taking your job in twenty-four hours!' And yes, he actually said 'Dude'.

'Give it a day and call me back.'

The next day, Johnny's friend had decided: 'No, I don't want it.'

That was all the encouragement Johnny needed. His boss signed off on his new posting right away. Two weeks later he was in Washington DC, then straight through to Indian Springs – later Creech Air Force Base – and Las Vegas. He joined the RAF's 1115 Flight, to which all British personnel were attached for administrative purposes. The designation '1115' came from a not-especially-creative combination of the 11th Squadron and 15th Squadron – the two USAF Predator squadrons where the British crews actually worked.

A short training course followed, lasting ten weeks. It included launching and landing Predators, basic handling and communications – all the usual flying elements, leading up to the firing of one Hellfire missile. Then straight to 15th Reconnaissance Squadron at Nellis Air Force Base, and operations over Iraq and Afghanistan. The most quirky element in the whole training process was the Brits not being allowed to fly in the northern part of the training airspace; it was too close to 'Area 51' and they might be tempted to look for ET.

At Nellis, the British contingent was made 'very, very welcome'. This might have been because they were such enthusiastic volunteers. Their American colleagues regarded it as an 'Alpha' tour, which meant they were taken off front-line flying for a tour on the Predator, at the end of which they *should* get the posting of their choice... but rarely did.

Over the subsequent months and years, the Predator would grow in importance as both an ISR asset and, being armed with two Hellfire missiles, as a ground-attack weapon. Even from the earliest days, however, there were indications that their geographical separation from the operational areas where the Predator was flown did not mean that the crews were emotionally distanced as well. More than a decade later, Johnny still recalls a number of events with great clarity:

'WHAT JUST HAPPENED?' – JOHNNY

There are a couple of incidents that occurred early in my time on the USAF Predator – 15th Reconnaissance

Squadron. You didn't even have to be in the GCS to be touched by events because everyone on the squadron could see the video feed from the Operations Centre.

On one occasion, a USAF crew was 'cleared hot' to strike a vehicle and its high-value target (HVT) with a Hellfire. They had followed the vehicle for some time until, eventually, it stopped at a point that looked ideal for the strike. The crew checked and double-checked the area, confirmed it as clear, and fired the missile. *Just* before impact – and too late to do anything about it – a kid on a bike rode straight into the blast zone.

'Bang.' Huge explosion. The screen blanked out and the kid disappeared. When the cloud of dust and explosive cleared, everyone watching was looking for the bike: the American crew inside their GCS, and everyone looking at the screen outside the GCS. The crew saw what happened – what they did – but so did everybody else who was watching the video feed. After all the post-strike assessments were made, the crew had to come out of their station and walk into the Predator Operations Centre. No one spoke.

What do you say? It was all within the RoE; the collateral damage assessment had been done. It was a terrible moment that couldn't be undone. That was in 2005, and it has stuck with me; I can still see the imagery now.

But the biggest thing that happened to me, personally, was an incident when I was the Mission Commander in the Operations Centre, overviewing the crew involved. An American helicopter had been shot down, so the Quick Reaction Force was sent out on a rescue mission in a Bradley Armoured Personnel Carrier. They were just north of Baghdad.

The Bradley pulled up at a junction and several of the soldiers got out and walked around in front of it, using mine detectors to make sure the road was safe to drive across. Once they were satisfied that it was safe, the men got back in the Bradley and it drove straight over the junction. It was followed by a second Bradley right behind it. But, instead of perfectly following the line of the first vehicle across the junction, it was off-line by probably a foot.

Then 'Boom'. The explosion shocked everyone.

The screen completely blanked out because of the heat and dust of the explosion – it was at night as well. The back door of the vehicle was blown off. Everyone inside was killed. The vehicle was taken apart to its component pieces by the blast, and we had to watch this. It seemed to happen in tremendous slow motion. We just sat there initially thinking, *What just happened to the screen?*

Then we had to remain on-station throughout everything that happened afterwards: all of the radio calls; getting the Medevac team in; watching it all. It turned out that there had been an enormous propane cylinder full of explosives packed 6-7ft underground on this junction. It could have been there for months. And we don't know whether it was remote triggered or what. That was on Memorial Day in 2005.

For some years afterwards, my wife and I used to have a get-together with the crew that was on duty that day, to remember. That flash on the screen, and the feeling of impotence, just stayed with us. Our job was to provide overwatch on these guys, to protect them. We had been staring, looking for anything that might be a threat. But there was a big puddle over the junction, and the soldiers with their hand-held detectors couldn't see what was underneath. That whole incident has stuck with me. I'm not saying I have PTSD, but I'm saying that I get how some people are affected that way.

It was hard work. A lot of twenty-year-olds were seeing this kind of thing for the first time, totally unprepared, and on a work schedule back then that did not allow time to see a chaplain or take time off to decompress. They'd be instantly back in it the next day. It is also sometimes difficult for Air Force personnel to talk about mental trauma or PTSD in front of other military personnel, like the Marines, for example. I have seen someone write on a blog, 'I'm suffering a little bit,' only to receive a reply that says: 'Try picking your friend's ear up off the floor while you are walking round avoiding parts of him on the ground. Try picking his bones out of your teeth. Then come and talk to me about PTSD.' And they're absolutely right. You can't compare it. From the perspective of the Predator though, and later the Reaper, those are the kind of things that were just being worked out as we went along. We just got on with it.

Just 'getting on with it' alongside Johnny was Tim, who would go on to make the step from flying the Predator to flying the Reaper.

THE FLAG

In April 2004, Tim moved from the UK to Nellis Air Force Base in Nevada to be an SO on the Combined Joint Predator Task Force, for a three-year tour of duty. By the end of that time he was the Chief SO for the USAF's 15th Reconnaissance Squadron. During the third year of his tour, the USAF was in the process of setting up the 42nd Attack Squadron and equipping it with the new MQ-9 Reaper. Tim was invited to be the Leading SO on the new squadron. The word 'invited' should be interpreted in the military sense, rather than in the sense of having much choice.

Tim's time living in Las Vegas was extended, and from 2006 he helped to set up the 42nd Attack Squadron from Day 1. His involvement would be good practice for what was to follow. Also in 2006, the UK decided it was going to 'stand up' (acquire) its own Reaper squadron, and Tim was one of those whose experience flying the Predator ensured that he would be involved in the process. Before any facilities could be built for the planned RAF Reaper squadron – No. 39 Squadron – someone had to decide where it would be located. Not just in the general sense of 'somewhere at Creech Air Force Base' but, more specifically, the exact bit of desert it would occupy. Someone would have to walk out into a suitable, available open space and mark it out.

Tim described the process in all its complexity. 'It was late Spring 2006, and a typically hot desert day. We literally walked out at Creech Air Force Base and into the desert with the Union Jack, found what we thought was a good area and said, 'We're going to have our squadron here!' The two of us planted the Union Jack in the ground and started setting everything up from that point.' It was probably the last-known occasion where a Briton has planted a flag to claim (or borrow) a piece of foreign land. They returned from the chosen piece of desert scrubland to see the Base Commander and begged him to allow the RAF to locate their squadron there.

They got the OK, the only proviso being that the Union Flag had to fly lower than the American flag. A metaphor, perhaps, for the past 100 years of British-American relations.

While Tim and the other British personnel started setting up 39 Squadron, they were still doing operational flying on the

USAF 42nd Attack Squadron. In addition, in 2006 the first group of British military personnel – not just from the RAF but from the other armed services as well – who would form the initial core of 39 Squadron, arrived in the US to do their Reaper training.

The plans for the squadron building were drawn up, based on specifications from the *ad hoc* planning team. It would be a prefabricated construction, made from units that almost 'clipped' together. The number of units to be combined to create a functioning squadron building depended on two things: knowing the future operational requirements of the squadron, and understanding the mentality of those in charge of the budget.

On the practical side, considerations included: the amount of crew room space that was needed for the number of personnel involved; the Ops Room; a room for digital networks and data links; coffee bar; and an outside wooden decking area where people could relax between sorties. In addition, the building and wooden decking area also needed to have wheelchair access, a legal requirement for all new American buildings. To the front of the building there would be a concreted area where all the GCSs would eventually be located. It would also hold further data links, and the hardware and software necessary to ensure the distant Reapers could be successfully flown remotely from that location. On the psychological side, the planners on the ground decided to play clever. Demonstrating the kind of entrepreneurial flair that made America great – and the kind of disregard for convention that is common among aircrew – the team asked for 150 per cent of what they really needed. Their calculation was that even after big reductions they would get exactly what they wanted. And they did. Everybody was happy.

With the first major obstacles overcome, the delivery and construction of the squadron facilities did not always go exactly to plan. Once the main building was assembled, a local company was due to deliver and fit the budget-price carpets. The van from the carpet firm arrived at the main gate to the base with two men inside. The guards commenced their usual security checks, then a delay ensued.

Meanwhile, back at the fledgling 39 Squadron building, Tim and a more senior colleague were waiting for those carpet contractors. Eventually, a phone call came though from the guards at the main gate.

'Two men have just arrived to deliver your carpets. Unfortunately, one of them has been identified as an illegal immigrant and arrested. Can you come and help the remaining individual to transport the carpet to your squadron, because he can't do it on his own?'

After one quick check of the ranks of the two available RAF people – Tim, the Master Aircrew (non-commissioned), and a Squadron Leader, his senior – Tim found himself in the carpet-moving business. For Queen and country. The birth of 39 Squadron may have depended on the development of technology that existed only in the realms of science fiction a few years previously. But it also needed the kind of buccaneering, can-do mentality that could see a senior SO go from lifting carpets from a truck to carrying out a Reaper training mission over California, and then conducting combat operations over Afghanistan for the USAF 42nd Attack Squadron, all in the space of several hours.

OVERWATCH

In the midst of the rapid developments and frenetic activity that surrounded the preparations for a new squadron over those months in late 2006, Tim had an experience – a moment of total clarity – that defined his understanding of this new technology.

The event shaped his attitude to using the Reaper. He still recalls driving home to Las Vegas, realising that he had achieved something special for a large group of people.

His Reaper had been tasked to provide an armed overwatch to a Canadian Forward Operating Base (FOB) in Afghanistan. The FOB had been attacked every night for a prolonged number of days, which had gradually exhausted the physical endurance of the soldiers on the ground as they defended themselves around the clock. The Reaper could deploy missiles at a moment's notice – and they were simply tasked to fly above the Canadians. Hour after hour, Tim's job was to control the cameras and other sensing equipment as the crew scanned for threats or attacks from the Taliban. If actual fighting broke out, and the coalition troops on the ground had to fight to survive, then he would laser-guide weapons onto the targets. When the Reaper arrived overhead, they were able to use the radio to check in with the Comms Officer in the FOB. He provided the crew with information about the directions from which the previous attacks had come and what the Taliban's tactics had been.

Methodically, Tim and his colleagues spent over six hours searching all of the approach routes for encroaching threats, from miles out right up to the perimeter of the FOB. The Comms Officer continued to point out the nature of the attacks and describe the different effects the Taliban weapons had, for example, on outside walls.

The chat and the searching continued for the whole of the overwatch, getting on for seven hours. At the end of the night as the mission was concluding, it was absolutely quiet and, as far as the Reaper team could see, nothing was going on. No threats, no bad guys. The Comms Officer on the ground apologised for bringing out the Reaper when nothing had happened. Maybe the Taliban were aware of the Reaper's presence overhead, or maybe it was just coincidence. And then he added: 'Just to let you know, I am the only person who is awake on this FOB. Everyone else is getting some sleep and it's the first sleep they've had in days.'

Tim could not stop thinking about those words as he drove home, and the effect his Reaper had on the guys on the ground. Not through the actual use of weapons, but just because of the sense of security or protection it gave. The Reaper was uniquely able to loiter overhead for many, many hours, observe in great detail and attack with precision. Tim recalls, 'I have been part of numerous kinetic (weapon) attacks when I was part of the squadron, but that incident tops all of them.'

In fact part of the Reaper's sensing capability was discovered almost by accident, and it was a discovery that shaped much of the future use of the aircraft.

By late 2006 and into 2007, the first British Reaper crews had been trained by the USAF – augmented by the RAF exchange personnel – and were developing their skills as part of the 42nd Attack Squadron. The purpose of such a squadron is clear from the name: attack. It would use the Reaper MQ-9 or Predator B as a Close Air Support platform – a 'bomb truck', essentially. It would orbit overhead in a designated battle space, to be called in for missile or bomb strikes to support troops on the ground. The designation, 42nd, was reconstituted from a squadron lineage that can be traced back to 1917, when it became part of the United States Army Air Service, long before the USAF was founded in 1947. It would also be the first Attack Squadron in the USAF, and was officially activated on 9 November 2006.[15]

For the RAF personnel, including Tim, who had come from an ISR background and were still working out how best to use the Reaper, only using it to fire weapons would limit its potential.

---

15 New Attack Squadron announced: http://www.nellis.af.mil/News/Article/286079/first-attack-squadron-stands-up-at-creech-afb/, accessed 10 January 2018.

It made sense to see what could be done with all the sensing equipment.

One of the first mysteries they faced was confirming whether or not the Reaper had a SAR on board.[16] If so, they wanted to start experimenting with the radar to see what it could do. SAR is designed to create high definition two- or three-dimensional images, especially of landscapes, while the parent aircraft flies overhead. This could be particularly useful while searching for objects or places in the Afghan countryside.

In response to informal enquiries, Tim was told – though without a high degree of certainty on the part of the person who told him – that there was no SAR on board. Not being convinced, he went to the Reaper hangar and persuaded an engineer to take off one of the panels and have a look inside. And there it was. On his next sortie, in full test aircrew mode, Tim proposed to switch on the radar, having devised some flight reference cards for doing that. While activating the SAR in a Reaper over Afghanistan, an American colleague ran into the GCS and told them not to switch it on.

'Why not?' He wondered if there was a fault with it. Maybe it would interfere with other essential systems on the aircraft.

'If you turn the SAR on, people will know that we've got that surveillance capability and they'll want us to be an ISR platform.' 42nd Squadron was keeping the focus on its designated Attack

---

16 The RAF website sets out the different sensing capabilities of the Reaper, which include: 'an infrared (IR) sensor, a colour/monochrome daylight electro-optical (EO) TV and an image-intensified TV. The video from each of the imaging sensors can be viewed as separate video streams or fused with the IR sensor video. The laser rangefinder/designator provides the capability to precisely designate targets for laser-guided munitions. Reaper also has Synthetic Aperture Radar (SAR) and Ground Moving Target Indicator (GMTI) to provide an all-weather capability. Reaper can also provide geographic location information to commanders on the ground or to other systems capable of employing Global Positioning System (GPS) guided weapons. The aircraft is also equipped with a colour nose camera, generally used by the pilot to assist in flight control and during take off/landing.' https://www.raf.mod.uk/equipment/reaper.cfm, accessed 12 December 2017.

role. Exploring the capabilities of the SAR would need to wait. For now.

39 SQUADRON – FIRST FLIGHTS

By October 2007, the RAF had received its first Reaper – ZZ200 – from General Atomics. The only discernible difference from the American version was that it did not have the USAF markings. Nor did it bear the traditional RAF bi-colour roundel and RAF insignia. In keeping with the spirit of the times – when everyone was asked to do jobs beyond their specialisation – the Squadron Boss researched what markings the aircraft should have. He then arranged with the paint shop at RAF Waddington back in the UK to have the decals (or transfers) made up and sent to Creech. He then personally stuck the RAF markings onto Reaper ZZ200.

The RAF had bought the platform – the aircraft – but still did not have the GCS from where it would be flown. And the squadron's building was not finished. So the RAF was flying its Reaper out of a borrowed American GCS, which comprised two rooms in a building as opposed to the more familiar, mobile metal containers that would arrive later. The rooms belonged to the USAF Test and Evaluation Centre and also prompted one of the most positive, if serendipitous, developments of the UK Reaper Force: the location of the MIC in the same space as the pilot and SO.

The American approach was to have only the pilot and SO in the GCS, with all the MICs in a central intelligence hub connected via intercom to their designated Reaper crews. However, with all the security sensitivities involved, the British MICs could not be placed there. The simple solution: put the MIC in with the pilot and SO. Initially, that just meant putting in a video feed, secure comms and screen, plus a 'repeater' screen so they could review information from the sensor equipment. And it actually worked incredibly well. While the pilot and SO could make an instant assessment of what was going on, the MIC gave them an extra pair of eyes, an extra brain and a longer-term intelligence picture of the Afghan conflict.

To begin with, before the British mobile GCSs arrived, the UK's two rooms led into one control centre that belonged to the Test and Evaluation Group. The rooms also had the advantage of being within walking distance – by which I mean hot desert walking distance – of the main 42nd Squadron building.

The 42nd were flying operations during the day in theatre; their development was many months ahead of the British. The RAF managed to get its one aircraft into theatre, although without some of the extra equipment needed to make it work. So 39 Squadron at Creech got the night shift – going in around 2200hrs – to get to grips with the technology and start developing the Reaper's capabilities. They would initially fly only a six-hour mission, which they then regarded as wasteful. However, they would get to try out their kit during daytime in Afghanistan, and could at least see something when they tested the cameras.

The first successful 39 Squadron mission was flown on 22 October 2007, but that was actually Night 3. The first *attempted* missions took place on 20 and 21 October. The more formal description of what happened is that the GCS at Creech did not achieve 'connectivity to the aircraft'. The Reaper in question had been launched by the LRE, and was in the air over Afghanistan. The Squadron Boss at that time describes how it all came down to one switch that had not been set properly. As a result, the control signal that should have been beamed to Afghanistan via cable and satellite to take control of the Reaper didn't work.

> We had a switch mis-set and we couldn't work it out. Even with all the Pred[ator] experience in and around the GCS, we couldn't work it out, we couldn't grab it [i.e. take control of the aircraft via satellite link]. So it got airborne and the MCE (Mission Control Element) in Creech never got control of it. It was the third night before we got control of it.
>
> One of the first missions over Afghanistan was watching two guys digging by a road for six hours. All they did was dig by the road. They could have been planting an improvised explosive device (IED). We didn't think they were, but all we did was watch them for six hours. When we reviewed the tapes afterwards, we could tell within five minutes that they were not doing anything "bad", in a security sense. All they were doing was irrigating their field at night, by diverting the water channel. When we applied our wider understanding of how the Taliban planted IEDs – the process they used – we could tell when something did not fit. So it was a huge piece of luck that the MIC ended up in the GCS with the pilot and SO, and the benefit was immediate.

The squadron did quite a few weeks like that: experimenting and familiarising the crews with the equipment, not getting home until the middle of the night or early morning. At that point they were not a declared asset to ISAF, and were a UK-only platform, working for Task Force Helmand. They had a dedicated Squadron Leader Intelligence Officer based in the CAOC, organising where and how they would operate over Afghanistan.

All of that initial activity happened before Christmas 2007. Tim sums it up: 'So we were doing some good, basic stuff –

unarmed ISR. There was a bit of an embarrassment factor for me in that I was flying operations on ISR missions for the RAF, and teaching Close Air Support – weapon use – to the Americans on their training ranges over in California: some of the best training you can get. Live blokes on the ground you could talk to; real soldiers prepping for Iraq and Afghanistan, with proper JTACs to train'.

The irony that British Reaper crews – whose own Reaper did not have weapons – instructed American crews in how to deliver weapons in support of ground troops, would continue into 2008. Despite this limitation, the technology continued to demonstrate its value even when just being used for surveillance.

'ARE YOU TRAINED TO USE THOSE WEAPONS?'

Doug experienced a particularly interesting mission on 24 December 2007 in support of UK forces in western Afghanistan. He had been a Nimrod maritime surveillance navigator instructor before he became a Reaper SO, the ideal background when it came to seeking out needle-in-a-haystack, or Russian-submarine-in-an-ocean, targets.

A UK Reaper (unarmed) and a US Reaper (armed) sat in three-dimensional boxes next to each other. That is, they were flying in adjacent designated airspaces. They were on the same mission, supporting coalition troops on the ground and also supporting some Apache attack helicopters. Doug spotted one of the Taliban targets, noted his position and put a call out over the radio. Nobody was listening. So he 'flashed' the infrared laser target marker on him, which got the JTAC's attention. The next thing he knew, he was 'lasing', using a laser to mark the target for an Apache helicopter. This, from an ISR platform without a weapon.

The Apache pilot responded with a Hellfire. So Doug had guided a Hellfire missile onto the Taliban target. Intelligence confirmed, both before and after, that he was the correct target – Doug killed the right man. But nevertheless, it was something of a surprise. The correct radio response would have been 'spot' or 'capture', something like that. Not 'smash'. The British Reaper crew proceeded to cue targets for the American Reaper for the rest of the day, until the American Reaper was 'cleaned off', with all of its weapons used, also known as 'going Winchester'. It was an early demonstration of why the British Reaper should have been armed in the first place. For Doug, *two* armed Reapers would have been even more effective against 'a massive set of insurgents', ensuring that some did not escape to regroup.

That kind of experience was one of the reasons why the UK Reapers ended up being armed in May 2008. In the process of acquiring, fitting and deploying Hellfire missiles and GBU 500lb bombs, Doug received an email from a headquarters-based Wing Commander in the UK, asking if he and his colleagues were trained to use the weapons. Doug's response reveals just a hint of the frustration that sometimes – just sometimes – builds up between those on front-line operations and those who are further away from the action: 'No, I just thought I'd give it a fucking go, what do you think? I've been teaching people close air support since last year and now you are asking if we're trained. And, by the way, it's a bit late since we are getting the weapons next week.' How much of that made it into the official email reply is lost in the mists of time, but the question still rankles all these years later. In fairness to the Wing Commander who posed the question, it was probably difficult for him to conceptualise the front line in Afghanistan being located several thousand miles away to the west in Nevada. Ten years later, there are still people, military and civilian, who are trying to make sense of it.

Bringing in the weapons to 39 Squadron was a bit strange for those involved. They were already firing weapons within the American Reaper programme, but had to come up with processes that fitted in with RAF procedures and conventions. For many decades, by the time a new RAF aircraft was deployed on front-line operations, it will have undergone years of tests and trials, with all of the training manuals and guidance robustly in place. With the Reaper being acquired as an Urgent Operational Requirement for Afghanistan, these procedural elements were bypassed and had to be 'retro-fitted'. The overarching guidelines were that they had to be safe and effective. The physical location of the MIC was merely one difference to be taken into account.

As 39 Squadron made the guidelines up, and they did make them up, they had the American technical rules and RAF guidance – known as Group Air Staff Orders – for RAF platforms, and amalgamated them. The Americans provided the aircraft and sensing equipment performance data. The squadron got hold of some Tornado fast jet guidelines, which covered things like crew duty terms and conditions, because they needed to follow practices for RAF – not USAF – crew duty.

They looked at how Tornado squadrons were crewed, and how they conducted authorisations. On the Tornado Force, both pilots and weapon systems operators were able to fulfil the 'authoriser' role and oversee missions on behalf of the Squadron Boss. So the new Reaper squadron did the same: they would have pilots and SOs doing authorisations. 39 Squadron maintained another convention: they would only have commissioned officers in the authoriser role.

Initially, they had to rely heavily on the squadron personnel who had weapons experience from the fast jet world, and those who brought weapons experience from the USAF Predator and

Reaper programmes. The weapon-firing rate, initially, was very low – around one to two shots per month. There was also only one mission per day, and they were several hours shorter than they would eventually become.

MISSING CONVOY

Once 39 Squadron had its first airframe, it found out everything it could from the ISR experts in the UK at 5 Squadron, who operated the Sentinel airborne stand-off radar platform, about how to use the SAR. A couple of the Reaper operators went back to the UK and spent some time finding out how it worked and what its capabilities could be. In a paradoxical move, the UK initially bought the Reaper to be an ISR platform, while the 42nd USAF squadron they were working with were using the same aircraft solely as an Attack platform. As Doug eloquently put it: 'The Brits buying an unarmed version of an attack platform was, for want of a better term, a "mind fuck" for the Americans.' Perhaps not how senior commanders or MoD policy-makers might have described the dilemma of how best to use this new capability.

Tim loved it though: 'Within months, we were finding more targets using the SAR than our coalition partners, who were just relying on conventional intelligence.' As word spread, and the full potential of the Reaper began to emerge, the 39 Squadron trainers then went back to the USAF squadrons to teach them how to use the SAR. An early incident showed the potential of the capability.

The task was quite straightforward: watching some compounds and observing the pattern of life (POL). A message reached the crew on the secure internet-style chat that a large UK convoy had gone missing; it had not arrived at its destination at the planned time. Its last known position, where it had its last contact with the control room, had been several miles from where the Reaper was now orbiting.

The crew could not take their eyes off the main compound – or rather, they could not move the camera that was trained on the main compound below. Continuity of observation was crucial for gathering evidence and intelligence. The crew discussed the options, which came back to the SAR, which they were still learning to use.

They decided that they would use the SAR, without leaving their location, to see if they could find anything. So, slowly and methodically, they began to scan the area where the convoy was most likely to be. Quite a few miles away they managed to pick up an image of the convoy on the SAR, which finds things that are still rather than moving. Its reproduction of the contours of geographical features is so accurate, so detailed, that artificial objects like vehicles stand out.

Tim and the crew could see the long convoy had stopped on a road. They reasoned that the loss of communication had been caused by the use of electronic-jamming equipment to prevent the detonation of remote-controlled IEDs, or similar devices; it would sometimes interfere with regular comms. The Reaper now had two tasks: observe the compound with one system and watch the convoy – in an entirely separate place – with another.

As they continued to observe the convoy, up ahead they could see two cars. It was the middle of the night, pitch black. The cars were static but facing the convoy, about a mile away. The Reaper team kept trying to communicate directly with the convoy and also communicate with the headquarters control room to get *them* to pass a message to the convoy.

Eventually, by repeatedly scanning with the SAR, the crew noticed that the two cars in front started to drive out into the scrubland before turning and parallel tracking the road towards the convoy. It could have been a potential ambush by low-level, unarmed enemy, 'dicking', or something else. It was certainly unusual behaviour.

By this point, basic comms had been established with the convoy, which was definitely at risk. A commander, somewhere, decided that protecting the convoy was more important that continuing the compound observation, and the Reaper was sent to the convoy's location. Tim then gave the convoy commander the precise location of the suspicious vehicles and he sent some soldiers, equipped with night-vision goggles, to arrest the 'dickers'.

Tim later heard that the dickers were involved with a large IED. On the night in question, about a third of a mile in front of where the convoy had stopped – in a culvert running under the road – they found one of the largest IEDs used in Afghanistan up to that point. So the Reaper crew had managed to prevent quite a large incident and a potentially devastating loss of life. And it all came from initially not knowing what the SAR could do and 'playing around with it, and coming up with some great results'.

Tim's words might be an oversimplification of the work that went into the understanding and developing of how best to use the Reaper, but they capture the imagination and innovation that went into making it work. From the Squadron Boss to everyone else involved with the Reaper – and, previously, those who flew the Predator – those early pioneers set the direction for what would follow in terms of how the Reaper Force would operate in the future. That would eventually include a second UK Reaper squadron at RAF Waddington in 2012. Like all pioneers they could see the huge potential of the Reaper in military terms, but theirs was just the start of a process of evolution that would continue long after their lives and careers took them in different directions.

CHAPTER 5

CIVCAS

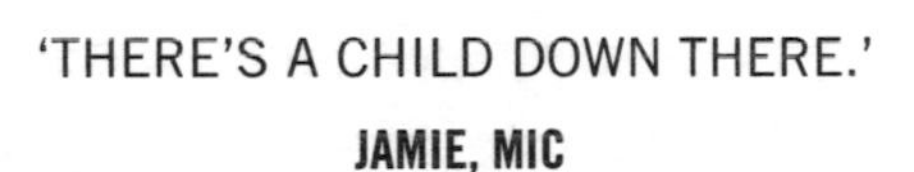

'THERE'S A CHILD DOWN THERE.'

JAMIE, MIC

The email came out of the blue, taking me by surprise. More specifically, two sentences took me by surprise.

'I would like to talk to you about the 2011 incident. I feel it is a part of the Reaper story that needs to be covered.' I re-read the words several times, digesting the enormity of the decision. Part of me was pleased that such an important story was to be shared. The rest of me was ashamed about being pleased.

I had begun to know Jamie over a number of months, starting with an interview in the autumn of 2016. My first impressions of him were still clear in my mind: serious, intensely analytical and extremely demanding of himself. I sensed something else in him that first time we spoke. There is no scientific or academic name for what I sensed, and it wasn't unique to him. It wasn't sadness or sorrow. It was something deeper that I had encountered many times in people that had seen – or delivered – death up close.

Something that I shared. I call it a soul shadow: a kind of darkness that changes its bearer forever and which never goes away. Like a black hole in cosmology, it is impossible to see directly. But you know it is there.

After the end of that first interview, once the recorder was switched off, he told me that he had been involved in the 2011 CIVCAS (civilian casualty) incident. I knew all about that because of the press coverage. I also knew that I could never have asked him to speak about it.

As he sat in front of me again, this time with the recorder on, he spoke slowly, haltingly. 'I couldn't have said before the incident how it would feel to know you'd killed women and children, accidently or otherwise. I always thought I would take myself out of the equation straight away and never want to be involved again. When it actually happened, that wasn't my reaction. I didn't have that thought process. And that's why I think it is difficult to try and explain. Why would I want to even risk going through that again? How can I live with myself knowing that it had actually happened?'

***

25 MARCH 2011

The pre-flight briefing had looked and sounded like every other. The MET report was good. Clear visibility with no cloud cover. A general round-up of important coalition activities in the operational area – Helmand Province – over the past twenty-four hours then followed. Crews were allocated to the distant Reaper airframes that were being prepped for take-off half a world away in Afghanistan.

Jamie nodded to his counterpart MIC in the flex crew that would relieve them for breaks and meals in the course of the next eight or ten hours. As the newest member of the crews, he was paying close attention to everything that was said, jotting down anything that might be helpful. His pilot – the captain – and the SO were making their own notes. Once he took his seat in the MIC station at the back of the GCS he would have to confirm that all of the intelligence they had just been given was still up-to-date and valid.

Specific information about the day's mission was minimal. They were given an operating area to go to. Further information would be provided when they got there. That could potentially mean a number of things: carrying out routine reconnaissance of the area; providing close air support to troops on the ground when they came under attack; or providing overwatch for some foot patrol to try and spot any threats before they got dangerous. They just might be needed urgently, so a shorter briefing would let them get their Reaper to where it was required. Not knowing what might transpire on a day-to-day basis was one of the things that Jamie enjoyed most about this still-new job on 39 Squadron.

If he had a choice, overwatch was his favourite activity. He liked the idea of protecting the soldiers on the ground. And for good reason. Not many of his colleagues on the Reaper Force had experienced life out on the ground during the war in Afghanistan. The danger, the consequences of things going wrong – sometimes with fatal results – and the dust were all still clear in his memory. He knew only too well what it felt like to come under Taliban attack and need air support to come to the rescue. He now liked the idea of being able to do that for others; it was a key reason for trying to get onto the Reaper Force in the first place.

He followed the pilot and SO into the box. Jamie still had not acquired their relaxed familiarity in the GCS. It would come with time but on this day he had only been combat ready for two weeks. Despite a couple of promising build-ups during that time, he hadn't yet been involved in a missile or bomb strike.

As he set up his MIC station he could hear the pilot and SO doing the same. They were getting ready to take control of the Reaper in mid-air from the LRE, who at that moment were launching it from Kandahar for its subsequent climb and mid-air handover.

The voice of the JTAC filled their headsets: 'We need you to go to this grid and match eyes with this other aircraft [i.e. train your Reaper camera on a target or area that is already being watched by another surveillance aircraft]. We're looking to prosecute when you get there.' As he read out the eight-digit map location in the Nowzad District of North Helmand, the three crew members all noted it down.

The other aircraft – which can be either manned or unmanned – can then undertake another tasking, fly back to base or stay on station to help with a complex scenario. To find out which it would be, and what they were flying into, Jamie started to engage with his Supported Unit to get their tasking and latest intelligence. By phone and internet-style secure chatrooms, he started to build up a picture of everything that was going on in that area. As he did so he could hear the pilot receiving the aircraft handover from the LRE with the words, 'I have control'. The only indication that the pilot had plotted a new heading was the changing digital direction indicator on the screen.

Their target area was best known for its opium production – an economic staple going back well over a century – and its stunning mountainous backdrop. The main town of Nawzad stood around 4,000ft above sea level, higher than almost every mountain in Scotland. It had also been strongly contested by the Taliban for years.

The handover and transit to the operating area took about

half an hour. It turned out that the other aircraft had been on task for more than eight hours and was getting ready to hand over responsibility to this fresh Reaper crew. Jamie gets the grid reference, gets the other aircraft's FMV feed and, with the SO, matches up the picture that he can see with what the departing aircraft can see. Quick and straightforward. A standard handover followed: they were now watching two trucks laden with home-made explosives (HME), with one male in each. One was a known Taliban commander and the two of them were moving about two tons of explosives further along a bomb-making chain that would produce and emplace a series of IEDs. In what seemed like another existence, Jamie had seen the effects of IEDs close up. Human devastation that couldn't care whether its victims were soldiers or civilians, adults or children. This was a priority target.

As the pilot went through the process of confirming the 9-line authorisation to strike, the sensor camera continued to track the two vehicles. As they watched from above, the trucks both stopped and one male emerged from each vehicle. They exchanged a few words, got back in and drove off.

This is going to happen, thought Jamie. After being combat ready for a whole two weeks and part of the Reaper world for all of two months, he was about to be involved in his first strike. His first lethal strike. Destroying tons of explosive and the men who would use those explosives to kill others held no qualms for him.

What they were watching was the first stage in a sophisticated operation that would ultimately lead to the planting of IEDs on or next to roads, trails, compounds and army foot patrol routes. They would be placed anywhere that the Taliban thought enemy or enemy sympathisers could be killed or maimed: either outcome was to be welcomed, even if it meant blowing up some of their own people as well. The system was well established.

The Taliban used someone with expertise to make tons of raw explosives in some isolated countryside compound away from coalition forces. The explosives would then be transported to the populated areas with a Taliban presence. The next phase would involve a bomb-maker creating a functioning explosive device, before the devices were finally put in place. Basic IEDs would be detonated by foot or vehicle pressure, but the more sophisticated ones evolved from controlled detonation by wire to detonation by remote control.

If the explosives in those trucks reached their destination and were converted into IEDs, they would be laid for British and other soldiers, and random passers-by, within days.

'9-line confirmed,' announced the pilot.

As Jamie scanned the picture on his screen for suitable strike areas, they were aided by the route that the vehicle drivers had chosen. They were driving down a wadi, a ravine where rainwater from the mountains might run off on a rare occasion. The drivers obviously thought they were safe from prying eyes but the good moonlight didn't help them, and the infrared camera picked up the missing detail.

'No collateral concerns,' said Jamie. He and the SO had been scanning both the near and far distance and there was nobody to be seen. Their targets were in the middle of nowhere, which is the ideal 'somewhere' for a Reaper to attack.

The crew agreed that this was the ideal location for a strike. As the pilot reconfirmed the 9-line details with the JTAC, Jamie became aware that he was only hearing one half of the conversation. The radio comms was intermittent. Great. There were protocols for such an occasion: pick up the phone, which worked perfectly for the pilot who had the phone but meant he then had to relay information to the other two.

'JTAC is calling for GBUs,' relayed the pilot, 'one for each vehicle. It makes sense if we want to take out all of the HME as well.' The GBU-12 500lb laser-guided bombs would not have been the first choice if they were just hitting 'naked' vehicles. On this occasion they were hitting trucks full of explosives and they wanted to make sure everything was destroyed.

'Roger.'

'Roger.' The SO concurred, followed by Jamie.

They knew the drill but the pilot talked them through it anyway. He had the trigger but they all needed to agree. Jamie and the SO had identified a suitable striking point not far down the wadi. The terrain would be forcing the trucks to move slowly. They would hit the first truck with a GBU. Then, as the dust cloud cleared, they would repeat the process and hit the second truck. It would not be able to keep going forward and would be almost impossible to reverse up the wadi. There was a good chance that the driver of the second truck would jump out and run away if he thought he was about to be hit, but what counted was destroying the explosives.

'Coming round.' The pilot was manoeuvring the Reaper for launching the bomb: not too high and not too low.

Jamie's heart rate was climbing, quickly. He had been told what to expect and had been waiting for the physical reaction. Though the disrupted comms was altering the dynamic of the process, the revised protocol was working. Textbook scenario. Isolated, middle of nowhere. No collateral risk.

The SO had the first truck right in the middle of the crosshairs on his screen as the pilot counted down: 'Three… two… one… Stores.' A gentle pull on his trigger released the 500lb guided bomb. The laser-guidance system is no respecter of day or night. The infrared camera gave the picture on the screen an eerie hue.

'Fifteen seconds.' The clipped tone told its own story.

'Ten seconds.'

'Five... four... three... two... one... Splash.'

It was odd to see the blast but not feel the explosion. No thump. No shock wave. He watched the beginning of the impact plume just as the screen went white. The flash of light and heat temporarily overwhelmed the ability of the camera pod to make out features. It was designed to work at night with very little light. In the semi-darkness of the GCS, the crew's eyes reacted as well. The sudden burst of bright, white light in front of them made them blink rapidly as their eyes adjusted.

The billowing of dust and burning material from the explosion continued for what seemed to Jamie like an eternity. Thirty seconds... sixty, and counting. The effect lasted two or three times longer than it would have done with a Hellfire missile and its much smaller explosive payload.

The pilot continued with the plan to go round for the second vehicle, flying a giant race-track pattern in the sky to bring the Reaper back to the same angle he used for the first GBU. In the meantime the SO deftly adjusted his camera pod control to keep the billowing, whited-out image in the centre of the screens that sat in front of him and his two colleagues. His fingers had relaxed slightly but only for a few moments. His job was only half-done: he would shortly need to guide in a second bomb.

As the cloud of dust, heat and smoke dissipated, Jamie could begin to see the detail of what the GBU had done. A clear hit on the front of the vehicle. Even without full picture clarity it looked like it must have killed the driver. Since the 9-line authorisation was to hit both vehicles no further permissions were needed. The camera could pick up both trucks, with the second now static a few vehicle lengths behind the smouldering carcass of the first.

'Cleared hot.' The pilot was responding to the JTAC's clearance to take the second shot. As the Reaper was running in for the second GBU on the second truck, Jamie could still hear just half of the pilot's conversation.

A slight movement through the haze in the picture caught Jamie's attention and he leaned in slightly towards his screen, his eyes narrowing. He could just make out a man – he judged it to be a man from his height relative to the truck. He was pulling something out of the flames at the back of the first truck and tossing it to the side. It looked like he was throwing bags. If the man was rescuing unexploded HME from a burning truck he must be mad.

In the GCS the process for the second strike was coming to a head. As the pilot was about to ask 'Is everyone content?' – just before the three-two-one countdown to weapon release – the images clarify further.

Jamie's stomach lurches. Next to the man is a person half his size. 'Is this… is this… a CHILD?' For a split second he doubts what he is seeing. The picture on the screen does not match the picture he expects to see from the intelligence they have been given. What if he is wrong? Don't panic. But he could see what he could see, even if it didn't make sense.

'Abort. Abort.' He interrupted the pilot who was focused on the final few seconds of the weapon release, while the SO was tense, waiting to guide the next bomb.

'What's the problem?' asked the pilot, his grip on the trigger snapping loose at the same time. The release process stopped instantly, the pre-strike tension suddenly broken.

'There's a kid down there. A child down there.' Jamie only had the perspective of seeing one person versus another, but even with the limitations of the infrared picture, that comparison between the two figures was clear. He might not have noticed if the child was standing on his own.

All eyes were on the figures that emerged more clearly out

of the gloom as the seconds passed: the crew in the GCS; the Auth and other observers in the Ops Room next door; and who knows how many watchers in the CAOC. Two questions dominated: What is that child doing there? And why didn't we know about it?

The pilot eased the controls and the aircraft went into overwatch as he discussed options with the JTAC. As the camera angle changed with the orbit of the Reaper, more people started to appear below.

'Multiple pax [passengers] from the second vehicle.' The SO kept the camera trained on the scene, briefly panning out to get a wider view in case anyone else was entering or leaving the area. Unlikely, in the middle of nowhere, but not impossible.

Pax, perhaps three, from the second truck were now doing something at the back of the first. It did not look like they were pulling bodies out. Whatever they were pulling out – HME was still the most likely – they were throwing to the ground off to the side. The child still stood next to the man, the viewers' attention repeatedly drawn to the small figure.

Jamie passed all of the information to the Supported Unit, while the pilot spoke to the CAOC. The Auth and the SMIC in the Ops Room next door were already aware as they were watching the same video feed and listening to their comms. Jamie continued to watch the movements of the passengers from the second vehicle. After removing some objects from the back of the smouldering truck they began to walk back down the wadi in the direction they came from.

He typed his observations into the intel system. There was no rush, no apparent panic by the people below. Perhaps they did not even realise the truck had been hit from above. The noise of an old, straining engine would dull any sound that would have come from the descent of the bomb in the moments before impact. Did they think they'd hit an IED: an ironic possibility. Maybe they thought the explosive went off by itself.

The group straggled their way back up the track, eventually sitting down. They were identified as women and children. The driver of the second vehicle walked off in another direction. By this time a second aircraft was supporting the Reaper as it continued observing. The vehicle driver was in the crosshairs of the supporting aircraft and he did not rejoin the pax.

Despite the appearance of the unexpected civilians, the primary purpose of this operation was not yet completed: the HME in the second truck still posed a threat. Once the pax who had survived the initial strike had moved far enough away, it was deemed safe to engage the second truck with no further collateral risk.

Whatever analysis of events was going on elsewhere, for Jamie, the pilot and SO at the centre of events there was no immediate respite from the intensity of the situation. Perhaps it was intentional: keep them busy and focused. They were still in their seats and there was still a job to be done. They were tasked by the JTAC to engage the second truck and destroy its explosives. By now Jamie suspects that their observation and analysis of the second truck for this final attack is being augmented by multiple watchers in different locations. The pilot goes through the final checks and the others confirm that they are content with proceedings for this follow-up strike. One final look to ensure that the women and children are keeping their distance before they are 'Cleared hot'. Then the bomb is away.

It seems to take much longer to fall the same distance as the first bomb, despite the same number of seconds being counted down. For the SO, hitting the static target posed fewer problems. However, any opportunity for the crew to relax slightly with the static target was outweighed by the added pressure of what they knew would be a larger-than-usual unseen audience.

When it came, the second blast was even bigger than the first. The screen white-out lasted longer and the infrared sensor indicated that it was also burning hotter. Jamie could replay the video of the final fractions of a second before impact. The second bomb did not just hit the vehicle, it hit directly where the explosive was sitting. The impact must have initiated some additional explosive effect, the target going off like confetti-burst, spread over hundreds of feet.

As the after-effects of this second, larger blast ebbed into the darkness, word came through that a helicopter and troops were being despatched to the scene. The flex crew also arrived in the GCS to relieve them for their first break. Not much of a handover was needed. The incoming crew had been watching events unfolding in the Ops Room, and they were already on overwatch mode when Jamie and the others filed out into the chill of the Nevada night. It was almost freezing and the jolt of the cold briefly snapped him out of his overwhelming thoughts. Half a world away, a helicopter full of ground troops had been dispatched to the scene of the strike to assess the situation on the ground.

Word travels quickly on an air force squadron. By the time they got into the crew room – only a few yards away – everyone seemed aware that something had happened. The Auth wanted to see them all, and the SMIC wanted to speak to Jamie afterwards to check how they were and to see if they were fit and able to go back into the box at the end of their break.

It hadn't occurred to any of them that they might not go back and continue the shift, even Jamie, whose first weapon event had not gone as expected and which was already prompting questions: for him, for his fellow crew members and for the squadron. Now, after a quick round of hurried chats, terrible coffee and a trip to the toilet, they were back in the GCS. And by the time they had settled into their seats, full comms had been restored and the helicopter had dropped its soldiers about half a mile away from the site and several hundred feet from the surviving women and children. The refreshed crew were keeping a close watch on the friendlies on the ground. More specifically, they were keeping watch to make sure no unexpected threats emerged in this barren area.

Jamie and the others peered at every feature and changing shadow on their screens as the Reaper continued its orbit and the SO worked his magic on the camera ball. All the while, their attention kept going back to the unfolding scene below. From the way some of the soldiers were interacting with the survivors, it looked like they were treating injuries. Radio chat confirmed their assessments when they called for the injured to be evacuated to hospital (CASEVAC).

Back at the destroyed trucks, the crew's worst fears were realised as the dead were laid out one by one and covered with sheets of tarpaulin. At least one adult woman and one child could be identified easily. The crew could listen in to the comms from the ground troops and gradually built up a picture of what they had found: the fatalities and the injuries. What do you do with information like that?

It seemed wrong to just leave the bodies lying there. But that is what had to happen. Coalition forces learned from long experience that, somehow, the bodies would be taken and buried before sunrise. All that was left for the dead was an appropriate burial in accordance with their cultural and religious practices.

And so the overwatch continued. Round and round, hour after hour. Looking to see who turned up, what they did. Seeing the tarpaulin-covered bodies. Round and round, hour after hour, seeing the unmoving bodies.

Two things are going on in Jamie's head at the same time. The trained, instinctive part of his brain is still in mission mode: there is a task to be carried out and it had to be done properly. The other part of his brain intruded, thinking about what was happening on the ground, about the details he is hearing, and the after-actions and reviews to come. They would take care of themselves. All he could do was keep watch over the troops on the ground.

'Friendlies exfilling [exfiltrating],' Jamie said. After several hours on the ground, the last of the soldiers climbed aboard the helicopter that would take them back to the relative safety of their base. Their actions had seemed coordinated – deliberate – as they dealt with what was a known civilian casualty incident by then.

Soon after, a lone figure turned up, walking straight to the bodies. From his gait Jamie judged that he was probably an older man. He lifted the tarpaulin and looked for a good few seconds. What was he thinking? Was he praying? He bent down and lifted the body of one of the children – there were two – and walked slowly away. Perhaps he had been working out which body he could carry.

The final couple of hours of the shift passed without further incident. The normality of events immediately after they exited the GCS seemed abnormal. Every weapon release was debriefed after the event and the two bomb releases of the past few hours were no exception. Everything would be scrutinised, from the sequencing of events in the process to decision-making over what weapon to use. The discussion would be factual and clinical, supported by the evidence of the FMV feed. The CIVCAS element was briefly discussed in practical terms and then the crew were told to go home and get some rest. The larger part of the process would be picked up in greater detail tomorrow. There was no precedent for debriefing the specifics and implications of civilian deaths. The procedure that existed on paper had never had to be used before.

As Jamie headed to his car for the hour-long drive home, he was left alone with his thoughts. With a natural tendency towards self-reflection and internalisation of his feelings, his mood was sombre. On another day he would listen to the radio, but not today.

How did I find myself in this situation? I joined the Reaper Force to make a positive difference after the shit I experienced on the ground in Afghanistan. How did my first weapon event turn into a nightmare, an awful nightmare? What have I got myself into? Then a reality check: *What time do I need to pick up Jane and the kids from the barbecue?*

They had only been in the country for two months and Jane didn't have a car – he would have to go and pick her up from their friends' house.

I'm not ready to go and pick them up yet.

It was about 9 o'clock at night and, spotting a coffee shop, Jamie pulled in. He just needed a bit of space to put his cascading thoughts into some sort of order, to make sense of what had happened and to think about what was still to come. All the things he couldn't – wouldn't – think about when he was in the GCS still doing the job.

He replayed the events surrounding the first strike and its aftermath on a loop in his head with regular interruptions. *There will be some sort of investigation and inquiry.* He didn't know what that would entail but he had been told he would find out tomorrow. *What did I do?* He went through his actions again and again. *What could I have done differently?* With the information he had – nothing. *You can only see what you can see and you can only act on the information you have. We followed the process to the letter.*

As he finished the coffee the internal dialogue continued. *We saved the lives of women and children by aborting the second strike. But that doesn't bring back the dead from the first bomb.* More mental juggling. *We took innocent lives. We saved innocent lives. We took... We saved.* No amount of repetition made the moral arithmetic of the positives and the negatives balance out. But the repetition went on. *We took innocent lives. We saved innocent lives.*

After an hour it was time to go. *Get your smiley face on. And if they ask me about my day?* He would think of something. He paid his bill and tipped the waitress. She could have no idea what was going through his head. *I wonder if she can tell?*

He arrived after the kids' usual bedtime, so there was barely any small talk. Thanks all round; exhausted children determined to fight sleep for as long as they could. It wouldn't be the first time he had had to carry a sleeping child from the car to their bed. Images of an elderly Afghan carrying a child in the dark.

The Squadron Boss phoned later to see how he was and offer moral support: 'Get a good night's sleep – we'll talk through everything tomorrow.'

Despite the reassurances, for Jamie, sleep had come fitfully. An early morning call from the SMIC (who was also the Deputy Commanding Officer) didn't help. The SMIC asked how he was – what do you say in that situation? – and checked he was still coming in to work. He was.

In the kitchen, Jamie was lost in his thoughts as he tried to eat something. In the background, the morning news was like white noise filling the silence and he was aware of Jane as she made her own breakfast. She knew that something had happened, he had said as much last night before going to bed, although he had not given her any details. She could always tell.

As the newsreader continued, the words 'Afghanistan', 'airstrike' and 'civilian deaths' hit him. Jamie stopped chewing.

He watched Jane as she heard the words as well. He could see the realisation as she put two and two together.

'Was that referring to you?'

Jamie hesitated. 'Yeah, possibly, yeah. There's something going on.' He had not even received official confirmation of the civilian deaths, yet here it was on the news. Of course, he didn't need to receive official notification since he had watched it all happening. But, somehow, hearing it being announced made it more real. Perhaps the old man had gone to the local Mullah, or perhaps the survivors had told someone.

Whatever the source, the coalition headquarters had released a formal statement acknowledging the civilian deaths. This announcement hit Jamie hard. Events that still had not fully sunk in now became very real.

As he climbed into his car to head out to Creech and the inquiry process that lay ahead, Jamie knew that his world had changed, that *he* had changed. The Reaper Force had also changed.

Back in the privacy of one of the XIII Squadron rooms in 2017 I am struck by two things. The first is Jamie's lack of equivocation or self-pity. The second is the sense of immediacy of those events as he talks about them. I try to recall in detail some event from my life in 2011 but fail. If, though, I had thought about something every day for six years it would be firmly fixed in my memory.

I ask him to continue with the events of the day after the strike, when he went back to the squadron.

'We were questioned, obviously. I had to write an in-depth, blow-by-blow account of everything from prior to getting to work to the event being over. The transcripts of everything we'd said on internal comms and transmitted on radios [which are all recorded] were typed up, and all the MIRC [military internet] chats were printed out. It was all kept and basically sent off for the inquiry as you would expect. We weren't shelved but I think we were offered the chance – 'if you don't want to fly you don't have to' – while the inquiry was going through its process. I chose to carry on, to do what we do on a daily basis.' The inquiry would take three months in the end, adding another layer of strain.

I ask about his reasoning for keeping on flying at that time.

'It's difficult to explain.' He pauses. Whatever he is wrestling with has not been reduced to a simple script to be repeated *ad nauseam*. He keeps firm eye contact and I am careful not to break it.

'We [the crew] spoke the next day and we all carried on, because there was nothing *we* knew beforehand that could have prevented what happened. There was no malicious act. You know you did as required, as requested, as called for. In our minds we'd destroyed two tons of home-made explosive, which was going to kill civilians and going to potentially kill British and coalition forces. The whole idea of the strike was to prevent that.

'And then when the worst case scenario happens and we realise that there are people down there who aren't what you are expecting… The only consolation is that we prevented further loss of life.'

'Over a period, how did it touch you or affect you?'

'I had to think long and hard about whether I wanted to get back into the box again. It was a real question. To go through this scenario again would be unthinkable. You never forget the fact that every day could be that day. Every day that you go in the box, be it for an hour or for three or four hours or the whole day, you could end up in this scenario again. And that's for three-to-five years. You're taking on a lot of burden and a lot of risk. But you also know that your own actions were by the book.'

I struggle to find a civilian comparison that would exert the kind of life-or-death mental pressure of getting back into a GCS after a shot that killed civilians. Perhaps a surgeon faces some of that pressure after a bad operation.

'So how does the process affect you? Has it changed over the years?' Another lengthy pause. I am sure I can out-talk him by a word ratio of 10:1. Any stereotype of the gregarious extrovert aircrew could not be further from reality.

'I think my wife can always tell what I've been doing. Even now she says: "I can tell when something's happened, just from the way you are when you come home." I don't want to sound ghoulish but when you've had a "decent" strike event you go through a process of extreme adrenaline. It gets to a peak… you ride the crest, the strike happens, then you come down as the adrenaline subsides. The occasions when a strike doesn't happen – maybe because there are civilians around – is what causes you issues. You don't get that kind of pressure release when you have to leave a threat out there to cause harm or kill people. It sounds terrible but I probably get home a bit more buoyant than normal if I've had a decent strike that went well.

'That day, my first experience wasn't a good one. You wait for that point where you get challenged to do your job. That's what you are trained to do. And then it turns into a complete nightmare. It hadn't gone the way I thought it was going to be. That's why I had that *What have I got myself into?* thought.'

'And yet you continued doing that job for several years, took a break and came back again.' I asked that because I spend a great deal of my working life trying to avoid too much stress, so I find it difficult to understand what drives someone to actively and wholeheartedly embrace the level of pressure he is describing. And perhaps one of the reasons I am struggling to understand Jamie and his experience is that he is trying to understand it himself, even after six years. He continued, explaining: 'I wanted to continue because for me it was always for some greater good.

And degrading the enemy's capability was the primary role. Certainly not to injure or kill anybody that wasn't the enemy, and certainly not women and children, that's for sure. I never knew how I'd feel if that would happen, and again it's difficult to explain. You never forget. You never want to forget. But neither do I want to dwell on it, if that makes sense.

'People might say, "Oh you're disconnected from it", but you're not. You watch what happens, you see the after-effects of what happened – that day especially. You read the reports in the paper, the reports of what people said had occurred. There were photographs in the papers and such. I don't know if they were actual photographs of the injured people but they are women and kids. And it doesn't matter where you are from, what your background is... that never goes away.'

'So how do you look back on it now?'

'It took a long time to make a decision to accept what happened and try to move on. I definitely felt guilty. I distanced myself from my family. You *have* to go through a period of feeling bad about what's happened, of feeling bad about yourself. Asking, "What could have been done differently?" And you go through those processes over and over *and over* again. To then get to the realisation that there was no other possible outcome in those circumstances.'

Writing down these words I am unable to fully capture the hesitation, the inner wrestling going on in front of me. I am left with one burning question: 'After everything you went through, both you and your family, you know the consequences if things go wrong. So why did you come back to the Reaper Force?'

'I didn't walk away from this to live in shame forever over what happened. I think some people might have judged that decision quite harshly...'

Not for the first time it becomes clear that Jamie is extremely sensitive to the perceptions of others.

'I'm a 45 year-old guy and I know I can't do this forever. But this is the most important job I've ever done, maybe the most important job I will ever do. I get to make a real difference in the world.'

Our conversation ends and Jamie heads out of the door towards the Operations Room to find out what he needs to do next. The fight against IS has not paused or relented in the time we have been speaking. And he still carries with him those events of 2011 in this different theatre of conflict. But it goes deeper: it has been his daily incident, every year since it happened. He half-turns and gives me a wave from down the corridor. I sense that soul shadow more intensely than I did before.

POSTSCRIPT

Max was a key figure on 39 Squadron at that time, with particular responsibility for RAF Reaper tactics, and would go on to be one of the most influential members of the Reaper Force. He recalls: 'This CIVCAS incident prompted extensive reviews: squadron review; RAF review; legal review; coalition process review. The reviews and investigations extended up through the whole RAF command chain to the Permanent Joint Headquarters and to NATO headquarters in Kabul. All of the reports were provided to the UN Special Rapporteur. The investigations concluded that from the point where 39 Squadron became involved in the incident, until after the conclusion of the attack, the crew's actions were in accordance with the Law of Armed Conflict and their procedures and directives.'

CHAPTER 6

THREE MONTHS

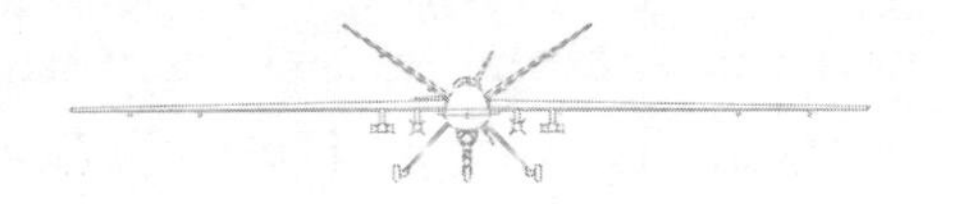

'PUT THE THING ON THE THING!'

TARA, WEAPON SYSTEMS OFFICER

'How are you feeling?' Marcus felt he should at least ask as the two of them orbited the kitchen. His Scottish accent did not lend a natural empathetic tone to the question.

The pause was just enough to acquire meaning.

'Fine.' Tara's reply did not obviously invite further enquiry as she finished getting ready to head out to work at XIII Squadron. She was already wearing her flying suit; living on the base had some advantages. Marcus, on the other hand, would have to change into his Royal Navy officer's uniform when he arrived at his work on a different Lincolnshire RAF station.

'Right then, have a good day.' Marcus wasn't sure what else to add as he briefly kissed her goodbye. It's not that he didn't want to ask more, or that he didn't care – just the opposite. It was Tara's sixth duty day in a row and he could see the fatigue.

He read the situation well enough to know that the best thing he could do was let her get on with it.

This is what passed for a normal day in the life of a couple who are married to each other as well as to two different Services: the RAF and the Royal Navy. It did not occur to them at that moment that Tara would make history by the end of the shift. It probably has not occurred to them since that day that her actions were unusual and deserving of a footnote in history.

More important was the mantra that they tried to follow as much as possible: 'Work is work, and home is home.' Those few minutes before they left for work and the few minutes after they got home were always a bit of a transition period. Most couples go through their own version of this process but the difference here is that very few husbands send their wives off to war on a daily basis.

Tara stood behind the relief SO for a minute, surveying the various screens before letting him out and easing herself back into the seat. So far, it had been another day of watching and waiting, looking for a human needle in a metaphorical haystack. She had been glad of the break. Her desire to empty her bladder had been slightly more urgent than usual.

In the months ahead it would take longer for her to get settled, taking an extra couple of seconds to set the seat slightly further from the desk and make sure her lower back was comfortable. Whoever had designed the SO's station probably had not thought about whether being pregnant would affect the control ergonomics. On this day, a few months into her pregnancy, there was not too much difference. In the coming months, the physical changes would happen so gradually that she would hardly notice the tiny adjustments she was making.

When she had first become aware of her pregnancy she had

let the Boss and the Squadron Executives know. The only early effect was fatigue and the only concession to that fatigue was being moved from the night shift. Most Reaper crew members have to fight tiredness in the early hours of the morning. Most people's circadian rhythm – or waking/sleeping cycle – hits a low point between 2am and 4am. Apart from that, she worked the same shifts as everyone else.

Tara was the squadron guinea pig as far as flying the Reaper while pregnant was concerned. Had she still been a Weapon Systems Officer in a Tornado fast jet she would have been grounded as soon as she reported her pregnancy. The Squadron Boss did what he and his predecessors had done with most new challenges since the Reaper came into RAF service. He used common sense.

Tara would keep flying as long as she was able and willing to do so. It turned out that she was able and willing until the eighth month. The Reaper proved to be a great equalizer in that regard: no aircraft manoeuvres exerting extreme forces on the body and no ejection seat to worry about. It didn't occur to her that she should be treated differently, and it made her feel guilty for a long time afterwards even to be moved from the night shift. While it reduced her fatigue levels to be moved off nights, she felt bad that others would pick up more night shifts in her absence.

As she resumed scanning the countryside looking for her target, there was no immediate indication that weapons might be involved on that day. There was also no concern or anxiety about the always-present possibility that things could change drastically and quickly. It had been several weeks since her last missile engagement, but part of the brain was always waiting for the signal to kick into that 'weapons' gear.

The pilot maintained the search pattern that covered two contrasting landscapes: flat, dusty plains with few major features, and the mountains that rose rapidly out of them in Eastern Afghanistan. Flashes of vegetation highlighted the places where there were water sources, often accompanied by isolated houses or hamlets.

Behind Tara and the pilot, the MIC was in constant motion, reassessing the video footage whenever there was anything that might be interesting or threatening. At that stage they were doing a routine scan, a 'search for badness' without a specific target, while also being available to be called in to support troops on the ground in any emergency. At the same time, he was chatting to his intelligence sources, both on the radio and via the secure military internet chatrooms. There was a high degree of confidence that the intelligence information they were working from was sound. That being the case, they just had to keep looking and hope for the best.

The biggest problem was the mountainous terrain. Huge crevices provided rivers of shadow that ran down from either mountain peaks or from the miles of long ridges between peaks. The 'river' effect in the shadows came courtesy of the heat in the air and the shimmering that the Reaper pod transmitted back to them. From within the major crevices, smaller creases in the rocky surface ebbed away like river estuaries, which they would have been when the ice-age glaciers that formed the terrain melted away. More shadows to hide in.

Dusty roads or trails through the mountains that were big enough to hold vehicles were reasonably easy to spot and follow. Tracks that were used by animals or people on foot were a different story. Even when short stretches were identified, they could easily trail off into nothing.

'Tara, can you zoom in on the shadowed area at the top of the screen?' The pilot brought the aircraft round to maintain the best angle to see the side of the mountain in question. With the expanded view, it was not a specific detail that had attracted his attention, it had simply been a movement within the shadow.

With the camera now focused on two people a ripple of anticipation spread through GCS. An initial assessment suggested they could well be of interest.

The crew broke into their well-rehearsed responsibilities. The JTAC had already started developing the 9-line authorisation and began to work through the various elements with the pilot. They were waiting on the crucial element, the positive identification (PID) of the target from the MIC. Were these men the ones the CAOC had been looking for? They were analysing still pictures from the video footage, and then confirmed the fine detail with the SMIC in the Ops Room next door and with the intelligence cell in the CAOC.

As well as working up the 9-line with the JTAC, the pilot had to keep the Reaper at an angle where the camera pod could see the target. The men were near the mouth of a cave on the lip of a mountainside. If the Reaper went behind the mountain they would lose line-of-sight and have to start the whole PID again.

The location of the men was posing problems for Tara. From certain angles the sun was almost blinding the camera for several seconds as the aircraft flew its orbit. Switching to infrared view did not improve matters hugely. Because the mountain rock had heated up considerably where it was under the direct glare of the sun, and the thermal effects coming from parts of the rocky mountainside reduced the clarity of the targets.

It only took a few more minutes to confirm the identity of the armed men. (Being armed in this area of Afghanistan did not automatically mean that they were Taliban fighters, hence the need for intelligence confirmation.) What had started out as a routine ISR mission was turning kinetic. With the 9-line and legal approvals signed off, the pilot started the final run-in for missile release. As the seconds wound down to the 'cleared hot' call from the JTAC, Tara sensed the onset of the adrenaline surge. She deliberately slowed her breath. If her pregnancy was changing her physiology, the effects were not yet noticeable.

After the pilot's familiar 'Three… two… one… Rifle,' and the firing of the missile, it was down to Tara. Her focus on keeping the crosshairs on the target was one of the few things this strike had in common with the first time she was involved in a missile firing, when she was flying in a Tornado GR4 ground attack jet. That, plus the gently increasing, involuntary bouncing sensation in her left leg, which she could not control.

Her first time had been 11 November 2010 – a highly memorable Remembrance Day – and her squadron had been in Afghanistan for just over a month on Operation HERRICK but, as yet, there had been no live weapon use. Tara was on standby alert duty in one of two Tornado GR4s that could be scrambled to provide ground close air support (GCAS). While the Tornados were required to be able to get airborne in thirty minutes, on this occasion it was close to the typical fifteen minutes that they aimed for. In an aircraft capable of exceeding the speed of sound there were few places in its operating area that could not be reached in twenty minutes or so. The mixture of weapons on board for typical GCAS missions[17] included two 500lb Raytheon Paveway IV laser/GPS guided bombs, three Brimstone missiles and up to 180 rounds of high-explosive incendiary ammunition for its 27mm cannon.

After the 'scramble' call came through there was a frantic rush to get the two Tornados in the air. Once airborne, the crews had a few minutes to calm down and get the aircraft set up for a transit to the mountain region a couple of hundred miles away.

---

17 https://www.raf.mod.uk/news/archive.cfm?storyid=2A5AA827-5056-A318-A87C9A1B7740085D, accessed 2 November 2017.

Initial attempts to use the secure radio to clarify their tasking were unsuccessful. Tara and her pilot were in the wingman position in the two-ship formation and were having difficulties chopping between radio frequencies as they progressed through different areas of communications coverage.

Once reliable radio comms were established they had to catch up with the details of an active 9-line and get 'eyes on' to the spot where the last report had placed the target: a vehicle that was being tracked by the lead aircraft in the pair. The transit flight, communication with the JTAC, locating and positively identifying the target, and getting authority to strike all took time. Once both Tornado crews had clarified their target, the lead aircraft broke off from the formation. It then linked up with a refuelling tanker to make sure that the pair would have maximum target coverage between them if the operation extended. For Tara and her pilot it was 'Yoyo' – You're on Your Own – time.

The pilot focused on getting them in the correct piece of airspace and keeping the aircraft in the right 'attitude', with the correct angle, speed and direction. Tara, as Weapon Systems Officer, watched the ground picture through the small black and white Litening 3 pod screen. Meanwhile the vehicle with its high value target was speeding along the rutted, dusty roads. It was emerging from one mountainous valley, crossing a plain and heading into another valley. It quickly became probable that they would be taking the shot. Soon.

When the final 'cleared hot' call came through, several things happened simultaneously. A huge adrenaline spike surged through Tara's tightly strapped-in body. One momentary thought flashed through her mind as she went through the final weapon checks and the pilot called 'Rifle': *Oh shit, this is really happening! Everything I've spent years training for, don't screw it up!* Then there was her leg. In what would be her personal 'tell' – a unique physical trait – her left leg started bobbing uncontrollably. Another thought: *What the hell is happening to my leg?* She had no control over it whatsoever.

Tara concentrated on keeping the crosshairs on the moving vehicle with its Taliban occupants. Another brief thought broke through: *Thank goodness it's not my right leg!* Her right foot would have activated the foot pedals for the radio. For the twenty or so seconds of the missile flight all of her available mental and physical capacity was concentrated on keeping a tight grip on the pod controls and keeping the crosshairs on the target. That intense focus ensured that the visual images of the missile hitting its target would be burned into her memory. There were no celebrations in the cockpit, only relief. Tara would learn later that there had been a more celebratory mood in the Operations Room of the ground unit she was supporting. They were watching on a live video feed to see days and weeks of preparatory work come to fruition with that strike.

Control over her left leg returned as quickly as it disappeared, coinciding with her relief at a successful strike. Somehow, a couple of Taliban fighters from the targeted vehicle had survived and were attempting to flee on foot. The lead Tornado had returned from refuelling to again lead the attack on the fighters. In the meantime, Tara's jet had to break off from the engagement and find the refuelling tanker.

By the time she returned to the strike area with a full fuel load, the engagement was over. The remainder of the sortie was uneventful but two related events would complete Tara's memories of the day. As the formation approached the base to land they radioed the control room, giving them the code that meant they had dropped weapons. She can still recall the excitement in the engineer's voice. It drove home the realisation that successful strikes from the air do not just belong to the people who pull triggers and keep crosshairs on targets, they belong to a whole infrastructure of people, expertise and equipment that make it happen.

The other event was the weapon debrief. Tara had not realised how long it would take, going over the minutiae of events that happened in the air at 420 knots, with less-than-perfect comms and only a small cockpit screen to work from. She would learn that all weapon debriefs are lengthy affairs, and that the amount of information available and recorded in a Reaper cockpit meant that they lasted just as long on the Reaper Force.

And back in the Reaper GCS, where Tara now worked, she could not afford to think about the debrief that would follow this strike. The missile was in the air, and on her screen she could see the effects of the ambient heat of the mountains and the bright sunlight on the camera pod. Despite the static target it was harder than she expected to keep the crosshairs where she wanted them.

Her left leg was beginning to bob, as she now half expected during a weapons event. Not as dramatically as when she had her first missile shot from the Tornado GR4 and it had appeared to hop about like it didn't belong to her. The leg bounce was also slightly less intense than it was during her previous, first shot from the Reaper. But it was enough for her to be aware of.

'Fifteen seconds,' called the pilot.

Time sometimes seems to change in the SO's seat when a missile or bomb is in the air. When a target is large, static and straightforward to hit, the seconds tick away as normal. When the target is moving, or there is a very fine margin for error with civilians at risk, or strange atmospherics prompt minute adjustments, time slows down.

As Tara controlled the missile between the '15 seconds' call and the start of the countdown five seconds before impact, those ten seconds felt stretched. The crosshairs created an imaginary micro-pattern as they drifted fractionally left and right, up and down. Not far enough to cause a miss, just sufficient to make the heart beat that bit faster. The nature of the target and the environment meant it was not advisable to 'grow a track' on it – that is, to use a degree of electronic assistance. So the missile was hand-manipulated for quick reaction.

She missed the small, ultra-sensitive control that she had used to direct the missiles from the back seat of the Tornado. Small inputs had been sufficient: tiny muscle flexions rather than distinct movements. The Reaper control felt like it needed great big arm motions in comparison, though in reality it was just more of a combined wrist and hand movement. A hefty tennis stroke compared to the light swish of a badminton racquet. Tara also missed the relative simplicity of 'putting the thing on the thing', as it was called in Tornado parlance. She was referring to the way that the crosshairs would digitally lock onto a target and the missile would automatically follow, all with the added benefit of not having electronic control signals delayed by being beamed across continents via satellite.

By the time the pilot called 'Five… four… three… two… one… Splash,' the crosshairs were right where she wanted them.

Relief. At least partially. BDA would still be needed to study and record the outcome. The effects of the heat and the sun on the stone and dust thrown into the air in the blast area were captured on the video feed, making it take slightly longer to see the images of the blast clearly. If the strike had been on an open, flat area, the pilot could have flown a simple circular orbit around it. But with a strike like this, the Reaper needed to fly a more elliptical, racetrack-style pattern that kept to the correct side of the mountain. That way, the camera – and the SO controlling it – could maintain line-of-sight.

As the dust settled the two dead were confirmed and the physical damage to the mountain was analysed. The Hellfire has only a fraction of the explosive capacity of 500lb, 2,000lb or even bigger bombs. But important lessons can still be learned about how mountains and explosives react in particular conditions.[18]

In the process of the after-strike actions, Tara had not even noticed the return of full control of her left leg. She was too preoccupied with everything else that was going on. Her camera work was as essential at that stage as it was in the preparation of the strike and her guidance of the missile. The effects of light and heat continued to slow things down by limiting the angles at which they could see the area clearly.

Time once more appeared to be elastic, speeding up markedly from the near-standstill of the last few seconds of the missile's flight. Before Tara knew it, the relief crew arrived to take over and let them out to go and eat. She eased out of her chair, stretching uncooperative leg and back muscles as she did so.

'Good shot,' whispered the replacement SO as she squeezed past in the confined space.

'Thanks.' Tara was confident it had been, but would reserve her final judgement until she had watched the footage again in the debriefing. Old habits.

She took the stairs back up to the Squadron Operations Room more deliberately than she used to. It didn't even occur to her that she had just made her own piece of RAF history, being pregnant and taking the fight to a distant enemy on the twenty-first century front line.

---

18 J. Underwood & P. Gluth (eds), *Military Geology in War and Peace* (Boulder, Geological Society of America 1998) explored a vast array of effects of different weapons in different places across history.

Everyone's journey to the Reaper Force is unique, as are the motivations for doing so. For those who move from flying other types of aircraft, a major factor is giving up the actual sensation of flying. The freedom to move in three dimensions in the air is particularly difficult to give up, at least where there is a choice. As a consequence of the 2010 Strategic Defence and Security Review, the Harrier fast jet fleet and the Nimrod maritime reconnaissance fleet were both retired from service. In subsequent years, all the remaining Tornado air defence fighter squadrons would follow, leaving just the growing Typhoon force and a diminishing number of Tornado ground attack squadrons as the RAF's combat force.

During Tara's recounting of the events that took her from being a Tornado GR4 Weapon Systems Officer – they used to be called navigators – to the Reaper, her tone ranges from wistful to enthusiastic. It came about as a result of a mixture of fortunate and unfortunate events. The unfortunate event that precipitated the major career change was the announcement, during an operational deployment in Afghanistan in December 2010, that her squadron was to close. That did not boost morale.

Tara's first major decision was whether to try and get a place on another Tornado squadron – which might also be decommissioned in the near future – or take the bigger step to Reaper. If she moved to another Tornado squadron and it, too, was closed down, she could find herself near the top of the 'cheap-to-make-redundant' pile during an era of RAF downsizing. Joining the Reaper Force would let her use her flying skills, albeit from the ground. The goal of a permanent commission and a full career would also become more achievable.

An influencing factor in that decision was how best to have some quality married life with Marcus, a Royal Navy officer. As the third decade of the twenty-first century approaches, the chances of the RAF hitting a small moving target with a missile on a far continent is very high. But the chances of two serving personnel, of any rank, who are partners being located in the same place for an extended number of years still falls into the realms of hope and luck. When it is a marriage between RAF and Navy officers, hopes of postings that provide domestic stability recede further. As someone who has conducted several military weddings in the past, I am acutely aware that they are often a triumph of hope over wisdom.

Tara went to the Reaper Force and was able to join the newly re-formed XIII Squadron in Lincolnshire in 2013, while Marcus managed, somehow, to get an exchange tour from the Navy to the RAF – also based in Lincolnshire. Happy days. Genuinely. In Lincolnshire. But before Marcus made it there, and while Tara was doing her Reaper SO training, he had to make a small detour. The kind of unplanned small detour that arises from time to time in the armed forces. In November 2013 he had been on the aircraft carrier HMS *Illustrious*, steaming back to the UK in time for Christmas, when Typhoon Haiyan hit the Philippines.

The order was given for the ship to go to the Philippines to support the humanitarian relief effort. Four days at 'max chat' (maximum speed) brought the ship to Singapore. Hundreds of tons of emergency supplies, food and water were loaded in twenty-four hours before the journey resumed. In the Philippines themselves, three weeks of manic aid work went on around the clock. As an air traffic controller, Marcus's job was to make sure the helicopters took off and landed safely on the ship. Air traffic controlling at a large airfield in Lincolnshire would never have the same excitement. But back to Tara…

She casually mentions that she flew Reaper operations while pregnant.

'How pregnant?' I ask.

'Up to eight months.'

'And what was it like flying the Reaper and using weapons while pregnant?'

A pause for reflection.

'I think some people may have thought I was a bit more hormonal, a bit more – not emotional – maybe a bit more angry.' The sentence finishes more like a question than a statement, as though she is exploring her own memory for the first time. She laughs at the recollection. 'Put it this way, the pregnancy didn't make it harder.'

It is too good a moment to let pass unexplored. '"Some people might have thought I was a *bit more hormonal...*" Would you agree with some of those people in retrospect?' She laughs again, neither confirming nor denying anything. Marcus, on the other hand, remembers those days well and is less reticent about offering his recollections.

'There were some days where you'd wake up and think... *I feel sorry for the person sitting next to her for the next six hours, because if they look at her the wrong way...*' The sentence is left hanging as he laughs nervously, half expecting that she is listening somewhere. He then quotes Tara: 'In her words, "It's difficult on this platform because you have to resist the urge to just punch the person [the pilot] that's next to you. Because in the Tornado fast jet you couldn't actually reach them. All you could do was turn their volume down if they were annoying you, but now they're actually sitting next to you."' I'm not sure even he knows the extent to which she was being serious or not. I'm not sure I can work it out either. Many a true word...

Tara is very open about her experiences, quietly proud not only of her operational flying with the Reaper but also in having been the guinea pig of the first woman to keep flying and deploying weapons through to an advanced stage of pregnancy.

'I definitely was more tired. They took me off the overnight shifts after a couple of weeks of the Execs knowing [about the pregnancy]. Before we got to the point of being able to tell other people, I just quietly stopped being programmed for night shifts after I fell asleep on duty with the Squadron Boss next to me.' She laughs again.

'It was a long blink, really – I wouldn't even say I woke up. It was one of those moments. Then the Boss's voice saying – loudly – "Tara, what are you doing?"'

'I'm fine sir, I'm fine.'

'No you're not. Right, somebody else is switching you out and you are going to lie down.'

'OK...' A thoughtful tone. Perhaps there is still a sense of regret that fatigue had overtaken her at that moment. 'They were very supportive throughout,' she adds.

I take the opportunity to further explore the ways in which her experiences as front-line aircrew challenges historical stereotypes of men and women when it comes to war. It is thirty years since the American political theorist Jean Bethke Elshtain wrote a book entitled *Women and War*, in which she challenged what she called the myths of 'Man' as the 'Just Warrior' who would protect the weak, vulnerable, non-combatant 'Woman' who is the 'Beautiful Soul'.[19] Yet Tara identifies herself strongly as the 'Just Warrior', who happened to be pregnant when she was killing enemy fighters with Hellfire missiles.

She makes her own comparison with a group of Kurdish women fighters of the YPJ – Women's Protection Units – who have been very successful in close-quarter conflict with IS jihadists.

By the time the RAF Reaper Force joined Operation SHADER in late 2014, IS jihadists had achieved a series of significant

19 Jean Bethke Elshtain, 'Women and War: Ten Years On', *Review of International Studies*, Vol. 24 (1998) 453.

advances as they sought to establish a new caliphate that would arc across Iraq, Syria and the wider region. The atrocities they committed against the Yazidi population in August 2014 brought worldwide media attention, notoriety and condemnation.

From early 2015, some of the most consistently effective military opposition to IS came from the Kurdish armed forces, with the YPG acquiring a fearsome reputation. Alongside the male battalions of the YPG fought a battalion of women – the YPJ. The success of the women in combat operations came down to physical bravery and a desire to confront a hated enemy. Every woman who took up arms in that way knew that they would face certain death if captured; the only delay to that death would be filled with some combination of slavery, torture and sexual abuse.

The success of the YPJ is due, in part, to their military prowess, accompanied by the psychological impact they have on their jihadist enemies. At the root of this psychological impact is a strand of jihadist, extremist theology that conventional Muslim scholars and leaders queue up to reject. For example, if IS fighters die in jihad they are promised entry to Paradise and an eternity with their God in the company of seventy-two virgins to help pass the never-ending time. But if they are killed by a woman then there is no Paradise, no virgins and no priapic eternity.

When the YPJ were defending places like Kobani and Rojava in Syria, they developed a distinct battle cry, a high-pitched scream identifying them as women. The IS jihadists faced the choice between standing and fighting the women, so risking eternal sexual deprivation, or running away and keeping open the possibility of a frisky afterlife.

What IS fighters all over Iraq and Syria might not appreciate is that many of their number have been excluded from their Paradise because they died at the hands of female Reaper pilots and SOs, like Tara. In addition, female MICs will have played crucial roles in positively identifying targets and potential targets, i.e deciding who gets killed.

There is therefore a great potential psychological impact on enemy IS fighters who fear they might not die in the 'correct' way. To take advantage of this, the roles of women in the Reaper Force fighting against IS could have been more widely publicised. It is a phenomenon that most, if not all, women Reaper crew members are aware of. Not only did Tara, a Reaper SO, raise the topic, so did Marcus, who waved her off to war in Afghanistan, Iraq or Syria every day for several years.

'If they knew who was dropping bombs on these guys it would absolutely tear them apart.' I assume he means metaphorically. Warming to the theme he continued: 'If the RAF could somehow, while protecting her and us and everybody else, release a picture of a pregnant woman dropping the bombs from the Reaper, it would have a huge impact.' However, to do that would mean treating the women crew members differently – singling them out for publicity purely on the basis of their gender. Perhaps a more general statement about both men and women being routinely involved in operations against IS might be better.

Tara also provides an insight into some of the differences between different aircrew cultures. That is, the cultures surrounding different kinds of aircraft, in particular the differences between the fast jet world and the Reaper world.

'As you go through training I don't think most people realise you are being pigeonholed into your own [flying] streams: multi-engined aircraft, helicopters or fast jet. You don't realise you are being taught certain ways until we all mix on the Reaper platform.' When aircrew talk of things being done differently on different types of aircraft, they usually mean one correct way – theirs, on the aircraft they flew in the past – and multiple wrong ways of doing thing, i.e. everyone else's way.

Tara provides an example. One practice that still has the ability to get her animated is radio communications.

'In the fast jet world you say as little as possible unless you have to.' That's because when you are flying at 420 knots things are happening at high speed and there is no time for waffling. 'A good cockpit's a quiet cockpit.' She expands on the point with enough of a laugh to suggest she has come to terms with it, but not quite enough to convince me. She describes her frustration when she started flying the Reaper with some of the pilots from multi-engined aircraft like the Nimrod or Hercules, as well as from the helicopter world. 'The guys would just *talk* and *talk* and *talk* and *talk* and *talk* and *talk* and *talk*.' Yes, that's seven 'talks'. With emphasis.

She continues: 'One pilot in particular is really difficult to fly with when it comes to the weapons thing, because he never shuts up. You stop listening to him because you're concentrating on your bit, thinking, *He'll pull the trigger in a minute... just get on with it!* Sometimes it is a bit more frustrating than others.' I sense there is more to be said but the conversation moves on. I am also sure that Reaper crew members from some of these different aircraft have a thing or two to say about flying with ex-fast jet mates.

Marcus's pride in what Tara has achieved shines through. He talks fondly of the extra 'daddy time' with the 'wee man' that he got when Tara went back to flying after maternity leave, and he took over many of the nursery drop-offs. He also recognised what she missed out on during the remainder of her time flying the Reaper. Meanwhile, the highlight of his time in the Navy – the HMS *Illustrious* disaster relief – is receding into the years 'BC', Before Child. Tara's Reaper experiences will go the same way. The ebb and flow of twenty-first century family life and military operations on the Reaper Force look similar to those in every other part of the military. Similar, but not quite the same.

CHAPTER 7

MISSING THE GORILLA

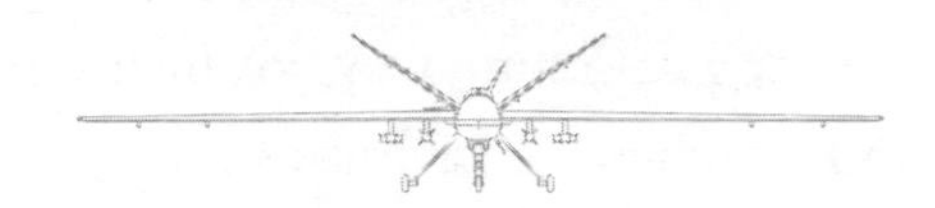

'BOSS, I'VE SCREWED UP. I NEED YOU TO COME IN.'

NINA, SQUADRON EXECUTIVE

The great fear of every Reaper crew is the 'bad shot' that inadvertently harms or kills civilians or friendly forces. A bad shot can be the result of a number of things. In the 2011 CIVCAS case, the processes were followed properly and the crew put the bomb exactly where it was meant to go, hitting the target vehicle. It became a bad shot because somewhere along the line – a long time before the Reaper even appeared on the scene – a vital bit of information was missed. To date, it is the deaths of those four civilians in Afghanistan in 2011 that casts the longest shadow over the Reaper Force. While that outcome is the one that causes the greatest anxiety amongst operators, there are varying degrees of bad shot.

A simple bad shot might be one that is taken from a poor angle when a better or easier shot would have been available with improved preparation or by waiting a bit longer. The problem there is one of process or technique, which will be subsequently scrutinised with the aim of making sure that others learn from it. After a private no-holds-barred debriefing with a duty instructor, the crew concerned will usually go through the learning points from the incident in front of the squadron at a morning briefing. From a squadron perspective, it is part of a continuous learning process. From an individual perspective, it is about owning up to mistakes and successes. It is part of a learning culture in the midst of ongoing operations; there is no time to sit back and relax. In a place where good performance both saves lives and takes lives in the most proficient way, all in the same day, or in the same few minutes, there is little time for sentiment or self-absorption.

The 'weapon event', i.e. missile or bomb use, that keeps some individuals waking up in cold sweats over a prolonged period, even after they have left the Reaper Force, is the *'What if...?'* shot. As in: What if such-and-such had happened? Or not happened? What if the explosion had been three seconds earlier or later? What if the blast had reached 3ft further?

One of the causes of the *What if...?* self-questioning is not unique to the Reaper Force or even to aircrew. Remember, where lethal weapons are involved the stakes are much higher. It is a psychological phenomenon known colloquially as 'Missing the Gorilla', based on a famous video experiment. Psychologists Daniel Simons and Christopher Chabris created a thought experiment called 'The Monkey Business Illusion',[20] which seeks to understand gaps in visual perception. The point of the experiment is to demonstrate how the brain can miss something that is blindingly obvious in retrospect, but which is initially overlooked because the individual is intensely focused on something else nearby. (Spoiler alert. The video clip is of a

20 The video can be seen at https://www.youtube.com/watch?v=IGQmdoK_ZfY.

basketball game where viewers fixated on the ball are oblivious to a man in a gorilla costume walking through the scene. Hence the title.)

The phenomenon has long been taught on aviation medicine courses. Taken to its extreme this is also known as target fixation. In the past this has led to pilots concentrating so hard on, say, a ground gunnery target that they have flown into it and died. The historian Peter Gray recounts such an incident from his early flying career. He was the navigator in the back of an F-4 Phantom fighter jet on an air-to-air gunnery exercise. He and his pilot successfully shot the towed banner that was their target, which then wrapped across the front of their aircraft as they flew into it.

On a UK Reaper squadron, additional layers of supervision have been added to prevent such events from happening. Crew resource management (CRM) – sharing out the responsibilities of a particular task between the three crew members – should stop everyone being fixated on just one thing. However, every so often even an experienced, intelligent, committed crew can get overly focused on a particular goal. Perhaps it is adrenaline and the fight-or-flight impulse that causes their focus to zoom in too far and prevents them from seeing the wider picture that is unfolding before them. Or perhaps it is just determination to see something through. Or a desire to stop someone murdering others. Whatever the reason, when it happens it can bring a *What if…?* moment that can haunt someone for a very long time…

Nina had the ideal background for a Reaper SO. Many years of development and experience had brought her here. She had always wanted to fly and got her foot on the first rung of that ladder by getting sponsored through university by the RAF. Her first flying experience came with her university air squadron, laying the foundation for a career as a navigator. She went straight though training and her first major personal success was to get her first choice of aircraft to fly: the Tornado F3 fast jet. The Tornado F3 was employed in an air defence role, which put a strong emphasis on air interception and air-to-air combat. Cheesy or not, the role is probably most famously associated with the film *Top Gun*. Eight years as a navigator was followed by a series of ground jobs typical of aircrew heading up the ranks.

There was one problem: the RAF retired Nina's Tornado F3 when the new Typhoon fighter came online. Unfortunately for her, the Typhoon was a single-seat jet with no need for a navigator. If she, like other aircrew, wanted a full career, she would have to be a Flight Commander somewhere to progress in the RAF. So, she would need to transfer to a different aircraft. The Tornado GR4 ground attack aircraft might have been one possibility, but that had also been around for a long time and might be phased out soon. To fly one aircraft type that gets retired is unfortunate; to fly two that get retired is just careless. ISTAR aircraft offered one potential avenue.

Looking to the future and where technology might take the RAF, as early as 2010 she saw the Reaper as the way forward: new technology, a new way of delivering air power and probably the start of a long-term shift towards remotely piloted aircraft. So Nina put her name forward.

She was pencilled in for a slot on the Reaper Force in 2012 but, in the way that happens in the military, another ground job came along. So it was in May 2014, after a five-year hiatus, that Nina stepped into a Reaper cockpit. Joining the well-trodden path of British Reaper pilots and SOs, she made her way to Holloman Air Force Base in New Mexico where USAF Reaper crews train.

Her background stood out at the course introductions where most of the pilots and SOs in the twelve trainee crews came with very little experience. She crewed with a British multi-engined aircraft pilot and there were a couple of other experienced aircrew making the transition. Even among the latter, her 1,000 hours of fast jet experience was unusual.

That experience served her well during the course. The very speed at which fast jets operate – often at supersonic velocity on the Tornado – forces their pilots and navigators to process large amounts of information very quickly, a definite advantage when they move to the Reaper flying at 120 knots. But that was all in the past. On this day, Nina had the responsibility of being the SO who would be guiding a Hellfire missile onto a moving vehicle in which an HVT was travelling. She was the senior ranking member of the crew, while at the same time being the one with the least Reaper experience. The pilot had carried out a similar engagement in the preceding weeks. As a Squadron Leader, it was also her turn on the rota to be the Squadron Duty Executive Officer with the Boss being away from the station. Plenty of eyes would be watching to see how this ex-fast jet navigator performed.

As the pilot transited the Reaper to the designated operating area, Nina confirmed as much information as possible with the MIC. She was aware of her adrenaline, which was still running at a low level. A number of things would now need to happen in fairly quick succession. They would first have to 'match eyes' with the ISR aircraft that was currently watching the HVT's car, as it moved as fast as possible on what passed for a main road through the desert. The first and most important requirement was to make sure that they identified the correct vehicle.

'Confirm eyes on.' The voice in her headset was calm. An operation like this was usually the culmination of days or weeks of painstaking intelligence gathering and surveillance by a number of ground-based and airborne sources. The higher the number of people who contributed to the overall effort, the higher the number of people who would be watching, or at least awaiting the outcome, from afar.

The initial identification was reasonably straightforward. There were hardly any vehicles on the road to cause confusion. Once the target was identified, Nina adjusted the camera aperture to get a closer look.

'Eyes on.'

'Roger.'

'Roger.' All three members of the crew confirmed that they could see the pick-up.

The JTAC, thousands of miles away in the CAOC, also confirmed that he was watching the same target vehicle. He began the 9-line approval process, working through its various elements with the pilot, the RCH, and his intelligence sources.

This would be a job for a Hellfire missile, not a 500lb guided bomb. They had a full complement of four missiles available. The pilot selected the one he would use and double-checked the missile settings. It was not as simple as pressing a button – he had to use the keyboard and type in his instructions. At the same time, he was talking to the JTAC and keeping the Reaper in the right part of the sky to be able to use the Hellfire.

The target vehicle managed to maintain a surprising speed as it kicked out a trail of dust. That speed transmitted itself to the supporting ISR aircraft and to Nina and her colleagues. As the MIC looked ahead to identify a clear kill zone, he was limited by the extent of the image available from the camera. Even with the supporting surveillance asset, they couldn't risk losing the vehicle because they would void the 9-line and they would have to start the PID process over again. The target vehicle was rapidly eating up the distance that was so vital in a successful strike.

Nina had to focus intensely to keep the target in the cross-hairs on her screen. Simple speed-time-distance arithmetic meant that if the car was doing 60mph, or 1 mile per minute, it would cover nearly 90ft per second. The satellite time delay on Nina's controls was greater than a second, so the vehicle would cover at least 90ft in the time it took for the camera or cross-hair controls to respond to her hand movements.

The control-lag took up more of her concentration. Even so, she fleetingly acknowledged a car that passed their target in a flash from the opposite direction. As their target sped along they also noticed it pass a couple of isolated buildings that weren't on the map.

The pilot was working to get the aircraft in the correct attitude – height, angle, speed, direction – for a strike. At the same time, the MIC was looking ahead for a suitable striking point.

They all spotted the next car that came down the road towards, then past, their target vehicle. But soon everyone is fixated on the target: Nina and the others in the crew, the JTAC and the supporting aircraft.

The next car to drive past is overlooked; a car containing unknown occupants – civilians.

Nina's adrenaline rises sharply, her grip tightening fractionally on her controls.

'Cleared hot.' The voice of the JTAC gives the final approval to the pilot.

'Everybody confirm happy.'

'Roger.'

'Roger.'

All three voices contain more tension than they did a minute ago. There is an intense focus on their target.

Another car passes by. The level of risk to civilians is rising – the missile impact point might not be clear if any of the passing cars slow down. A civilian car could end up being too close to the target vehicle.

'Counting down: Three... two... one... Rifle.'

A gentle squeeze on the trigger by the pilot.

Nothing.

The visual indicator of a Hellfire missile blasting off its rails towards the target should have been a slight wobble of the Reaper's camera pod and a fleeting distortion of the image on the screen. But nothing. Just an increasing number of cars on the road.

'Checking weapons.' The pilot immediately looks to the digital weapons readout on his main screen. It indicates that all the weapons are still on the aircraft.

Behind him and Nina, the MIC instantly rewinds the video feed by a few seconds to check for the lack of image distortion that goes with a successful firing. 'No indication.'

'Fifteen seconds.' All of this happens in a few seconds but the pilot still continues the countdown as if the weapon was in the air.

Nina keeps the crosshairs on the target as though the missile is heading groundwards – standard operating procedure for safety in this sort of incident. But she knows there is no missile.

Oh no. Nina is still keeping the crosshairs on the target but all she can now see are cars of civilians intermittently passing their target. The collateral risk had been too high. She looks to the safe 'shift' area to put the missile into, running the safety drill automatically. For the missile that is not in the air.

Several seconds after the missile should have exploded – if it had fired – Nina's queasy, sick feeling got stronger. As did the first internal questioning: *What if...?*

'Guys, we screwed up.' She didn't need to say it. They all knew. It was like coming out of a trance-like state brought on by the intense focus and getting caught up in the determination to hit the HVT.

They had Missed the Gorilla, big time.

The brain does interesting and unexpected things under pressure. Every football fan has watched their favourite striker standing in front of an open goal, waiting for the easiest of tap-ins, screaming for the ball. Except the team-mate with the ball does not see the easy option and attempts a much harder, ultimately unsuccessful shot on goal. TV coverage then tends to show the striker berating his teammate for being stupid, blind or a combination of both. Except the teammate may be neither of those things. They might just have been using the wrong part of their brain: the part that increases their chances of Missing the Gorilla.

There are several possible reasons as to why the brain fails to see things clearly in these extreme, high-pressure situations. One of the most influential of these explanations was developed by Daniel Kahneman and Amos Tversky, who wanted to understand the psychology of making choices and decisions. They identified two different ways of thinking: intuition (e.g. your instant impression of a politician you see on television) and reasoning (e.g. adding up your change in a shop). This approach to understanding how way the brain works was later described as System 1 and System 2.[21]

Most people rely on intuition a lot more than they realise. It is a kind of short-cut system that enables the brain to deal with huge amounts of inputs through sight, hearing, smell, taste or touch. The reasoning part of the brain will analyse some of those initial impressions to check if they are accurate. Under threat or under pressure, that reasoning part of the brain can slow down or even shut down, which leaves the instinctive, intuitive part at work.

One possible explanation for Nina and the rest of the crew

21 Stanovich and West (2000) proposed the neutral labels of System 1 and System 2.

Missing the Gorilla – being prepared to fire a missile while the risk to civilians was increasing – was that under pressure they started to rely increasingly on the intuitive part of the brain. Correspondingly, the reasoning side became increasingly less effective. So, the three crew members had slowly narrowed their field of vision and their breadth of thought to the target vehicle. It was probably over a period of less than a minute, but an eternity relative to the speed of brain signals. Their individual and collective intense focus was then broken by the failure of the missile to launch. Suddenly, the reasoning part of their brains was forced into action to identify what had happened, and why. The trained and well-practised process of firing the missile broke down. Familiar actions that they were taking intuitively suddenly did not correspond with expectations.

As soon as the no-fire was spotted, the collective spell was broken. The rational, analytical parts of their brains took over. The recognition of the error was instant and horrifying. *What if...?* Nina could not shift the tension in the pit of her stomach. The FMV feed of what had – or had not – happened had already been seen in real time by multiple audiences in the squadron Ops Room and the CAOC.

After a flurry of checks and hurried calls, in what seemed like no time at all, the relief crew arrived to take over. The walk to the Ops Room was nowhere near long enough. Very few words were exchanged as the trio climbed the stairs. As the pilot signed in at the Ops Desk, the Auth was waiting for them. The Flight Lieutenant Authorising Officer sat them down and spoke to the whole crew, together, before speaking to them separately. The situation was somewhat awkward – as a Squadron Leader, Nina was senior in rank, but in this situation the greater authority was delegated to the more junior Auth.

'I'm pulling your "Cat", right now.' With those words, the Auth revoked her Combat Ready flying category. 'Do you have anything you want to say?'

'No. You're absolutely right to do it. Completely. We screwed up.' Nina faced it head on. Part of that reaction was down to temperament; she was highly pragmatic about most things. The other influence on her response was conditioned. Having spent several years flying the Tornado, and even more years training to get there, the principle of no excuses had been drummed into her. Post-flight debriefings in that environment were blunt to the point of brutal. When tiny mistakes could cost you your life, either in air-to-air combat or just in flying the aircraft, eradicating them becomes an obsession.

The immediate consequence for Nina and the crew was that they could no longer be involved in weapon firing until they underwent remedial training. But since she was the senior officer involved, a Squadron Executive Officer and an experienced part of the team, all eyes would be on her. She had said the missile shot was on, and low risk. It wasn't. Her relative inexperience on the Reaper would not be seen as an excuse. It didn't cross her mind that there were any excuses to be made.

Nina's professional embarrassment was overshadowed by the deep knotted sensation that sat somewhere between her stomach and her rib cage. *What if...? What if...? What if...?* She had Missed the Gorilla. She had never put civilians at risk like that before. Now she had been relieved of her flying duty.

Remedial training would come in the following couple of days. There was a more immediate problem. Nina was the Squadron Duty Executive Officer and she could not now carry out that function. She would have to call the Squadron Commanding Officer and tell him what had happened so that he could replace her for the rest of the shift.

She returned to her office, took a deep breath and dialled the number.

'Boss, I've screwed up; I need you to come in.'

'What's happened?' No raised voice, a matter-of-fact tone. He would make a great poker player. Maybe Harrier pilots – and he was a highly experienced former Harrier pilot – really are as cool under pressure as people say. By 'people', I mean other Harrier pilots.

'I'll explain when you get here, but my "Cat" has been withdrawn.'

A slight pause. Nina wasn't avoiding the explanation, there's just a limit to what can be said about classified military operations on an insecure phone line. But she had said enough. If the problem involved her Combat Ready Category, the Boss knew it was a flying incident.

'Fatalities?'

'No.'

'On the way.' The Boss was relieved as he ended the call. He had been in the 39 Squadron building in 2011 at the time when the CIVCAS incident happened. The relief was only momentary. He still had to break it to his wife that he had to go back into work. She wouldn't be surprised.

The rest of the shift proceeded with much less drama but the incident had several immediate knock-on effects, beyond interrupting a rare few hours off for the Boss. The squadron was now a crew down. With the flex crew taking over as the duty crew, an additional crew would need to be activated to be the replacement flex crew. Three people who had not expected to fly today would now be getting the call. And that was just the beginning of the individual and administrative impact.

Then there was the flying rota, which existed in a constant state of tension between increasing demands for Reaper's use and the increasing difficulty of meeting those demands. The Reaper had proven to be highly successful over a number of years in its twin roles of ISR and attack. Foot patrols on the ground, from the regular British Army to coalition allies to special forces, wanted as much overwatch as they could get. Senior commanders also wanted as much intelligence information as possible – the lifeblood of every military campaign since time immemorial.

The result was a constant push to attract and train the right number of crews. This type of incident – where crews were quite regularly getting called into work from their precious days off – was a morale crusher. It was one of the reasons why only a minority of pilots, SOs or MICs volunteered for back-to-back tours flying the Reaper. There were two aspects to it. The first was the fatigue that built up in people over time as they flip-flopped between war and peace every day. The second was the negative impact on family life for many of those involved. A familiar military story, perhaps, but one that has a new dimension on the Reaper Force where combat operations are continuous and ongoing for years, instead of in four- or six-month blocks.

Other factors also influenced manning levels. The Reaper Force had benefited from its inception from using experienced aircrew from other aircraft types like the Harrier, Tornado, Nimrod, Hercules and different types of helicopter. That previous experience brought a disadvantage as well. Many were coming towards the end of the contracted length of their service, so they would not be staying for the long term.

The flying rota manager was facing the difficult task of rejigging the complex crew-flying programme. More ominously, he would be faced with the unenviable task of having to let everyone affected by the changes know what they would now be doing. Until the next revision. With so many families directly connected to the squadron, there was a good chance that yet another birthday or school play or other event would be missed. The building blocks of family memories can be either positive or negative. Either way, someone always remembers long after the latest squadron emergency has passed.

After a night being haunted by the *What if…?* question, coming back into work the next day was almost a relief for Nina. The feeling in the pit of her stomach had not gone away, fuelled by dented personal professional pride that she had not met her own standards let alone the squadron's standards. The sensation would magnify when she thought of what might have happened. She tried to treat the day, and other people, as normally as possible. But there was no escaping the awkwardness. She knew it; they knew it.

After the morning briefing and some routine admin, Nina and the rest of the crew assembled in the Briefing Room with the duty instructor for the standard debrief that happens after every weapon firing – or non-firing in this case.

The room is usually either empty or bustling with the activity of the daily briefing at the start of a shift. Its rows of empty seats provide an eerie backdrop when there are only four people present for a debriefing. It should be a place of cool, calm, rational scrutiny of the video of whatever operation or weapon strike is being analysed, and for the most part it is. However, the post-strike debrief can also be one of the most emotive and emotional parts of life on a Reaper squadron.

In the event of a 'vanilla' strike (the use of a missile or bomb that is routine, straightforward and a long way from any civilians), the debrief is usually short and uncomplicated. The debriefing officer, or senior non-commissioned officer, talks the crew through the video and through their written report of events: what was happening, what they were thinking and saying, how the strike happened and where there were margins for improvement. In the event of a complex strike, the dynamic could be very different.

The crew did not say much as they waited for the instructor to get their video up on the big screen. After every weapon event, and this non-event, the video recording, plus audio and chatroom messages of the few minutes surrounding the strike is captured from the live feed and archived. All videos are shared between the two RAF Reaper squadrons for formal training purposes as lessons are identified. Significant or sensitive videos make their way up the military hierarchy for further scrutiny. There is no hiding place from the permanent record.

They sat in silence during the first run-through of the video. Everyone winced as the HVT passed a steady but increasing trickle of potential civilian collateral damage – the occasional car on the road, a nearby building. Without the brain-altering effects of adrenaline, the real-time hyper-intense focus on the target and the pressure to get the missile on target with an international audience watching the live video, the misjudgement was glaring. The crew didn't need anyone to point out the error but it was going to be pointed out anyway.

The initial viewing was followed by a frame-by-frame analysis. Each point where there had been an opportunity to spot the increasing risk was identified and discussed. *What if…?* The error arose because of a combination of individual judgements and poor CRM. If the visual responsibilities had been shared out and followed more effectively – if the three crew members had been looking out for different risks and threats on the screen – someone would have spotted the potential problem earlier.

Watching the footage on a 6ft screen also magnified the errors – literally and metaphorically – compared to watching the 20in screen in the GCS. Reliving the experience was so intense that some of the crew almost thought the missile might actually fire this time.

Once the debrief was finished, a number of follow-up activities kicked in. The squadron was still carrying out full-on operations and they couldn't afford to be a crew down. Anyone who felt embarrassed or sorry for themselves would have to do it in their own time.

'You will all be on Remedial Action,' the instructor informed them. Nods all round.

A Remedial Action Report (RAR) consisted of a formal record of the error and a detailed plan of remedial training, which had to be completed to a sufficiently high standard to ensure that future operations could be conducted safely and legally. The report could then be closed. The RAR would remain on their personal files for years to come. A career would typically not survive more than one or two of those.

The remedial training took place over the next two days with the senior instructor. The problem had not been technical; the aircraft was flown correctly, radio and other procedures were followed, and the 9-line approval process was similarly correct. The problem had been one of poor judgement – individually and collectively. The senior instructor took the crew through a whole series of videos and scenarios that would test their judgement, reinforce why they went wrong and how to avoid doing it again. Their video would eventually become a feature of the training of new Reaper crew members: a how-not-to-do-it warning to others.

Once the senior instructor was satisfied that the lessons had been learned and had reported to the Boss, the three each had a 'check-ride' to confirm that they were ready to start operations again. For Nina and the others, it was a relief to get back to flying again. Before long they would each be involved in successful strikes, but not before one final contribution to the squadron training process. A kind of act of penance.

A few days later the Briefing Room was bustling as everyone settled down for the daily briefing. The usual items were conducted with customary briskness: intelligence update, weather in the operating area, the day's tasking, and so on. The Auth finished his part of the briefing with the words 'OC C Flight'. This was Nina's formal title as the officer responsible for the squadron's communications equipment, for the hardware and software necessary to make the Reaper work, and for Flight Safety.

Nina stepped forward to the front of the room as the Auth set up the video of the non-firing incident on the large screen. 'Ladies and gentlemen, for those of you who have not yet seen this video – this is how not to conduct a strike.'

Nina was the first person from XIII Squadron to volunteer to be interviewed on my first day at RAF Waddington. My initial impressions are still clear. The green flying suit indicated that she would be flying later that day and the single-winged brevet on her chest indicated, before she told me, that she had been a navigator in the past.

I am greeted with a cheery, 'Hi, I'm Nina.' She conveys a breezy self-confidence – common among fast jet or former fast jet aircrew. I don't know if she has ever played team sports, but I picture her as the team captain. As a father, I wish I could bottle that sense of purpose and self-assurance she exudes and prescribe it to the many British teenagers who seem so crippled with anxiety.

After a few pleasantries and introductions I explore with her some of the differences between flying a fast jet and a Reaper. There is no holding back.

'The human-machine interface is challenging – it's not intuitive and things are not laid out as you would expect them to be in a "normal" cockpit. Routine information you would need for flying the aircraft – heading, height, speed – are on different screens and can't be seen at the same time. But you learn how to use it and once you do, it's fine.'

She speaks warmly of her time doing her Reaper SO training at Holloman Air Force Base. She was a guest on the American-run course and training alongside some very young, very junior USAF personnel: 'They were awesome. We were students united by our experience.'

Given her background and previous practice, Nina expected to have a speaking role during her Reaper training, communicating with the JTAC. This turned out not to be the case, which became obvious the moment she touched the microphone.

'Sensors don't use the mic,' her instructor intervened.

'This one does.' The pilot was busy so she broadcast over the radio to help him. The US crews had more specific roles, whereas British crews all expected to have a speaking role. As an overall experience, after several years out of a cockpit, flying the Reaper seemed familiar, even though she had never previously had to do air-to-ground attack before.

The pivotal part of the interview came when I asked her to describe her most memorable moment or incident on the Reaper Force. When she could have told me about precise shots or crucial surveillance she had been involved in, she chose to talk about coming close to potential disaster, recounting the events captured above. That choice seems to have two dominant motivators, though there are probably other factors as well.

The pursuit of excellence is clearly important, in the way that many top sportspeople spend more time analysing and rectifying their errors than celebrating successes. The second factor is a kind of pride in the way that the operating environment and training system kicked into action when things went wrong, even when it was personally embarrassing: immediate suspension; analysis of events; Remedial Action; retraining; check ride; and return to operations.

Nina's determination to confront this particular failing head-on has not diminished in more than a year since the incident happened. She recalls what happened after she presented the video to her colleagues.

'I briefed them a couple of weeks later and watched everyone's face go white... I can never know what would have happened but we accepted far too much risk. We hadn't noticed that was happening and didn't stop when we should have done.'

She still wants others to learn from a potentially disastrous error in judgement to make them better aircrew. Perhaps there is an element of personal atonement there as well. At the end of our interview, she asks one of the instructors to show me the video and talk through what happened, or almost happened. That video is now used in the training of new Reaper crew members.

If Nina demands the same standards from other people that she demands from herself, I expect she will be a complete pain in the arse to work with, and a bigger pain in the arse to work *for*. Yet every great sports team there has ever been has at least one person like her.

Nina has not Missed the Gorilla again, but she still occasionally gets a visit from the *What if...?* question. She will never know if any civilians would have been harmed that day if the missile had actually fired.

CHAPTER 8

CHOOSING

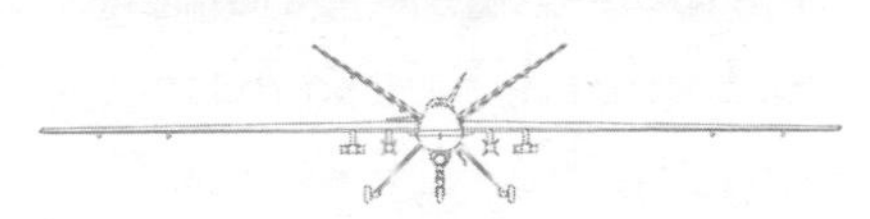

'WHICH ONE OF YOU IS GOING TO HAVE YOUR LIFE ENDED RIGHT NOW?'

JAKE, SENSOR OPERATOR

Since the First World War, military aircrew – fighter pilots in particular – have enjoyed a special, almost mythical status in Western culture. Standing above them all, metaphorically speaking, is the Red Baron – Captain Manfred von Richthofen – the German First World War ace officially credited with eighty aerial combat victories, or kills. One of the most decorated British aces was Major James McCudden VC, with fifty-seven kills. All aces shared many characteristics: bravery, skill, determination and probably their fair share of good fortune, at least until it ran out. The basic technology was inherently dangerous, the weather was often a threat and the odds of survival were poor.

The aces were also romanticised and stereotyped to a degree that eventually led to the greatest fighter pilot parody of all time:

Lord Flashheart, from the BBC comedy series *Blackadder Goes Forth* – arrogant, hyper-heterosexual, sophisticated, fearless and upper-class. The stereotype did not happen by accident: it was part of official attempts to bolster public support for the First World War through the liberal use of propaganda.

Lord Rothermere was the Air Minister on 1 April 1918, the day the RAF officially came into being. He wrote an article in the *Times* that day which would be ridiculed if it were written now, a century later. Entitled 'British Airmen's Daring', it waxed lyrical about the extreme bravery of 'the British flying man'.[22] He wrote about the success of British airmen against 'the Hun' being down to a combination of 'perfect physique, of matchless bravery, [and] of extraordinary quickness of brain'. These caricatures betrayed no inner wrangling and, presumably, the pilots were able to kill on a conscience-free whim. The survival of the fittest in noble aerial duels among gentlemen.

However, the reality for the aircrew was much harsher: shortened lives that experienced wild extremes of excitement and fear and cold, especially in the open cockpits. If they were lucky, death would be instantaneous – perhaps a shot through the head or heart. If they were unlucky, they would burn to death in aircraft made, essentially, from fire kindling. Or they might remain conscious as they plunged and hit the ground. Such realities ground down the best of them, including Britain's highest scoring ace of the conflict, Major Mick Mannock VC, who was haunted by the fear of being shot down in flames – which in fact turned out to be his fate.

Often forgotten was the youthfulness of those early pilots. On 3 May 1917, Captain Albert Ball (later awarded the Victoria Cross) wrote to his parents saying, 'I am feeling very old just

---

22 *The Times*, 1 April 1918, p. 8.

now'[23] when in fact he was two months short of his twenty-first birthday and Britain's leading ace at that time. He died four days later having shot down forty-four enemy aircraft. What was it that made him feel very old? Not just tired, but *old*.

How much was it living with the daily danger of being killed, and how much was it the killing? We will never really know, but he left clues. He also wrote to his parents: 'I like this job, but nerves do not last long, and you soon want a rest.' And of the men he killed: 'Nothing makes me feel more rotten than to see them go down.'[24] Brave – yes. Fearless – no. Happy with killing – no way.

The notion of noble duelling was more myth than reality. Early in the First World War there may have been occasions for single combat, through choice. But as the war wore on, massed air forces needed a different, less romantic, approach. McCudden captured the pragmatism of killing while avoiding being killed during an aerial attack on a German aircraft in January 1918.[25] He had been flying at 17,000ft, deep into German territory. Spotting his target far below, he idled his engine so that he could quietly glide down behind his prey, the sun behind him to help obscure him from being seen. He wrote:

> When I got within good close range, about 100yds, I pressed both triggers; my two guns responded well, and I saw pieces of three-plywood fall off the side of the Hun's fuselage. Then the L.V.G. [aircraft] went into a flat, right-hand spiral glide until it hit the ground a mass of flying wreckage... I hate to shoot the Hun down without him

---

23 Albert Ball quoted in Kennett, Lee, *The First Air War* (New York: Simon & Schuster, 1999), p. 170.

24 Chaz Bowyer, *For Valour: The Air VCs* (London: Grub Street, 1992).

25 James T. B. McCudden, *Flying Fury* (Newbury: Casemate, 2009), p. 253.

> seeing me, for although this method is in accordance with my doctrine, it is against what little sporting instincts I have left.

Note, first, he sneaked down on his enemy in a surprise attack and shot him from behind. Second, he was obeying his doctrine, his rules. Sporting? Noble? Hardly. Kill without risking being killed, the eternal military quest. And the Reaper is just the latest manifestation.

One question that I have harboured for many years about those so-called 'Knights of the Air' is how did they choose who to kill when there was more than one aircraft they could chase after? The nearest one? The slowest one? The one at the most convenient angle? Perhaps they just decided instinctively in the few seconds they had to choose. But there was still a choice to make. That's another area where Reaper crews have both an advantage and a disadvantage over their famous forebears: they have the time to choose. And they definitely have time to think about the choices they make.

In a series of events and personal recollections, Jake reflects on the hardest questions of all: Can I kill? Can I choose who lives and dies? Can I sit back and do nothing while watching jihadists commit murder?

It was 2014 and the Taliban target was walking along the side of a dust-strewn road. He could not have known that his image was at the centre of the screens in the XIII Squadron GCS, and on some live video feed at the Combined Air Operations Centre. If the man suspected that he might be targeted someday, the way he was moving did not give away any sign of urgency or panic.

In the GCS, the MIC had already confirmed the identity of the individual who was about to be struck by a Hellfire missile.

The pilot was confirming the 9-line authorisation with the JTAC and preparing the settings for the missile.

Jake, the SO, was keeping the image at the centre of the screens with his controls. His adrenaline was up and running, but not spiking through the roof like it would when he was trying to keep the cross-hairs on a fast-moving target. He had previously served in a mentally demanding role as a tactical navigator on the Nimrod maritime surveillance and anti-submarine aircraft. As a result, he could cope with the Reaper and had time to think about what was about to happen. He had previously guided missiles onto static, non-human targets so the process was not new. It was not yet, however, second nature.

He found that even basic communications between the members of the cockpit were more akin to those between Tornado fast jet crew members than between the crew on a Nimrod. For some reason – probably because so many ex-Nimrod personnel moved to the Reaper Force after the Nimrod fleet was disbanded – he expected a more familiar approach to comms. It wasn't that he was starting from scratch, just that he needed to unlearn some mannerisms and deeply embedded comms practices.

For example, on a Nimrod, with up to twelve crew members, when talking to someone else Jake would say their position followed by his own position. So, if he wanted to talk to the Radar Operator and he was the Tactical Navigator he would say, 'Radar-Tac Nav'. He would then wait for the Radar Operator to say, 'Go ahead Tac Nav,' before speaking. There was none of that formality in the Reaper GCS. He had seen new operators from the Nimrod take that approach on Day 1, then they would stop when they realised that wasn't quite how it worked.

Another thing that surprised him slightly was finding that he was now doing the same job as former fellow Nimrod crew members of different ranks and from different flying trades. While Jake was a Flight Lieutenant, his job was also being done by a Sergeant and by the Wing Commander Boss over at 39 Squadron. The Reaper is a great equalizer, especially with regard to the SO and MIC.

However, nobody comes, even from other aircraft, with all the necessary skills but many do come with varying degrees of the necessary basics that can be developed. People from a flying background, especially if they had highly mentally demanding jobs beforehand, are probably the best placed to make the transition. In Jake's case, because of his navigator experience, he instinctively understood airspace routings, tactical awareness and mission planning – the more conventional aspects of flying. These were skills that would later come in more useful when he was flying a Reaper over Iraq and Syria in congested airspace, rather than over Afghanistan with its relatively quiet skies.

In what seemed like no time at all, the pilot was 'cleared hot' by the JTAC, completed the final checks and countdown, then pulled the trigger.

'Rifle,' said the pilot. The missile was now rocketing through the air, and responsibility for hitting the target fell to Jake and his use of the laser-guidance system.

As those first few seconds of the missile flight time passed, Jake would later recall with great clarity what was going through his mind. *I'm going to kill someone. I'm about to kill someone.* He repeated it to himself two or three times as more seconds ticked away and his grip kept the crosshairs on the man going about his jihadist business below.

Then the warning… 'Encroaching civilians.' A second, supporting ISR aircraft spotted them and alerted Jake and his crew to the potential danger.

'Shift!' Jake and the pilot both called it at the same time.

The civilians were heading towards what would become the blast area in another fifteen seconds.

Jake reacted instantly with his controls. The adrenaline surged higher as he moved the cross-hairs away from the man walking along the road and towards the empty ground a few hundred feet away. The safe 'shift' area had been identified in advance. Just in case. With the decision made to abort the strike, the last thing he wanted was to kill or harm someone by accident.

The pilot counted down to impact: 'Five… four… three… two… one… Splash.'

As Jake's screen filled with dirt and dust from the explosion, he exhaled with relief. He had gone another few minutes without killing anyone.

Jake joined the Reaper Force in 2013, towards the end of operations in Afghanistan. He had just enjoyed a period studying the use of military force, and his own part in it. Spending time at university, away from the military environment, one thought dominated: *Can I kill?*

Jake applied to join the RAF back in 2005 and went through Officer and Aircrew Selection. He was asked in his interview: 'You are applying to join a fighting service. Are you prepared to take another person's life?' Like many people before him he instinctively answered 'Yes', (to say 'No' would mean failing), not fully grasping the implications. It was an abstract, almost surreal question.

As the months counted down to starting his training on the Reaper, Jake began to agonise, *Can I kill?* He had learned enough to know that the nature of Reaper operations meant it would be a matter of when, not if, he would be required to take life.

His time on the Nimrod had not prepared him for it. He had carried out training exercises in the Nimrod simulator where electronic torpedoes – no more real than blips on screens – had struck equally imaginary, enemy submarines. The external Exercise Controller would say, 'There's debris floating on the surface. There are bodies on the surface. Good kill.' He would then think, *Good job*, before heading off for a debrief that talked through a shot that did not really happen. The odds of firing a missile on an enemy for real seemed so slim as to be not worth seriously considering. If it happened for real it would probably be the Third World War, with the end of the world in sight. There would be other things to think about.

Despite all the deliberation, Jake was still not convinced he could take someone's life when he started his Reaper training. But he would give it a shot – as it were. He would throw himself into it and see what happened.

In early 2014, Jake had been on the Reaper Force for a year when he learned that he would be deploying to Afghanistan to join the LRE later in the year. By then, despite coming close on occasion, he had yet to kill anyone. He was starting to think he could get through his whole tour without having to do so. He began wondering, *Can I get to the end of this tour and not have to take a lethal shot?* The question was never far from the forefront of his mind during the subsequent months. *I might get away with it.*

But then in June 2014, just a couple of weeks before he was due to go out to the LRE for several months he took his first completed shot. Jake and the crew had been watching enemy forces in the Helmand Province of Afghanistan – an area of considerable Taliban activity – for several hours. The Taliban fighters were well armed, and it became apparent in those hours that their target was an American unit just over one mile away. Calm observation of general movements of the enemy fighters on the ground rapidly escalated when it became clear that they were starting to move into position for an attack.

Jake's crew targeted two Taliban on a motorbike. Authorisations took seconds rather than minutes. For Jake, once the missile was in the air, the responsibility was all his. The rapid development of events and the speed of the bike took all his concentration. He had no time to think about anything else. He had not wanted to get to this point where he would be taking the lives of others, but he did not shy away from it either.

The Reaper Force was not shooting many weapons at that point, so there would be considerable interest in his shot, not least from a 'continuous-training' perspective.[26]

The pilot counted down the final few seconds to impact as Jake hit the bike and its riders as precisely as he could have done. The screen lit up with the combination of fireball and dust.

Before the explosion could fully subside, one of the two men limped out from the fireball. On the infrared vision Jake could see where the blast had removed the lower part of his leg. Clear of the fire, the man collapsed. In the distance were passers-by who went to help him. There was no follow-up shot. The fighter was *hors-de-combat* – literally, out of the fight – and civilians were now surrounding him. Beyond the initial arrival of help on the ground, Jake and the crew learned no more about what happened to him. They still had a ground unit to protect, and they had to observe and report on the aftermath of the strike.

The other fighter on the bike was killed. Someone pointed out that it looked like a 'Hollywood shot'. A real-life version of Hollywood special effects and a large fireball, where the hero walks away unscathed. But Jake knew that the fighter did not escape unscathed. *I've grievously wounded someone. I haven't given him a swift death.*

---

26 Since the advent of the Reaper Force, crew members have had to do a considerable amount of learning and training on the job. Therefore, at that time every weapon event – missile or bomb – was shown and analysed during the squadron daily briefings so that everyone benefited.

Jake's reservations about using weapons had not stopped him from doing so. Despite the outcome of the shot – the wounding, not a quick death – Jake was satisfied that he had done everything possible: no negligence, no control errors, no unprofessionalism. The subsequent frame-by-frame reviews came to the same conclusion. So, having taken his first lethal shot, he left for his pre-deployment training in the US before deploying to Afghanistan to the LRE a few weeks later.

British forces were due to withdraw from Afghanistan by the end of 2014. Not long after Jake was deployed to the Reaper LRE in July 2014 there was talk about a possible move to a new theatre, Iraq. The Iraqi government was seeking international support to fight the advances of IS, which had been announced as a caliphate on 4 July 2014 by Abu Bakr al-Baghdadi, the self-declared caliph of that illegitimate, pseudo-state entity. Operation SHADER started in Iraq on 26 September 2014, with RAF surveillance flights extending to Syria within a month.

After two months in Afghanistan, Jake and the rest of the RAF LRE moved location to support Reaper operations in Iraq. It involved different local and regional politics, different terrain and a different, more overtly aggressive enemy. It took a couple of weeks to get everything in place – a testament to the efforts of logisticians whose work usually goes unnoticed and unappreciated in the background of military or humanitarian affairs.

Jake loved the new flying environment. The complexity of multiple coalition partners flying in congested airspace and setting up systems to make sure that everyone stayed safely in one piece brought his navigator skills back to the fore. Three months passed in a blur – on top of the time he had spent in Afghanistan – with no thought of having to use weapons. Inevitably, he found himself heading back to the UK and XIII Squadron at Waddington.

He returned having had a close-up view of how Operation SHADER against IS differed from his previous Reaper flying in Afghanistan. Gone was the emphasis on supporting friendly troops on the ground with ISR, and with only the occasional use of Hellfire missiles and guided bombs. Sporadic attacks against friendly forces by Taliban fighters in Afghanistan had given way to highly intensive IS fighting that needed to be disrupted. ISR was still central to Reaper activity, only now it was also feeding into the much-increased use of missiles and bombs in the Reaper's 'attack' role against IS jihadists on the ground. Every day of the first few months that British Reapers were operating over Iraq, Jake landed them after taking over the remote control from crews at either Creech or Waddington. The most important piece of safety information he needed every day was how many weapons the aircraft still had on board. While in previous months in Afghanistan it was unusual to land a Reaper that had had one of its weapons fired, on Operation SHADER it was the opposite.

So Jake was returning to Mission Control Element (MCE) flying: regular operations where he would be employing missiles and bombs again. His extended time away had not resolved his ambivalence about using weapons for lethal strikes. In total, between his preparation for Afghanistan, his time away and a leave period after he got back, ten months had passed since his one and only lethal shot so far.

After such an extended time away from flying full operational missions, he returned on a downgraded personal flying category. After an initial check, like everyone else who has any kind of significant break, he returned with an 'Op 2, Limited Combat Ready' flying category. He could fly most parts of the mission, but he would not be allowed to fire weapons until he had gone through the full 'work up' training programme.

After his first limited-role refresher flight, he spoke to his

Flight Commander and was candid with him. 'I've been away from this kind of flying for a while. I'm not certain about how I'm going to react or going to feel getting back into regular operations.' On his initial sortie, Jake was the SO for the mission and carried out all the preparation stages for a strike. Being Limited Combat Ready, he could do some things but could not control the weapon onto the target: 'stunt flying', or 'stunt sensoring' as it is sometimes called. So he got out of his seat and the Duty Instructor replaced him.

'He takes the shot and I think about six or seven people died. That was my first day back in the seat,' Jake recalled.

A few days later the Squadron Boss approached Jake and asked, 'Are you OK about everything?'

Jake assumed that his Flight Commander had spoken to the Boss and mentioned that Jake still had a few reservations about firing weapons.

'Have you spoken to my Flight Commander?' asked Jake,

'No, I haven't, 'the Boss replied, 'I'm referring to the shot you took just before you left.'

Jake's guarded response had been enough to alert the Boss to the fact that Jake's reservations were still alive and kicking, more than just a legacy of the fireball shot from ten months before. Jake was thinking, *I'm not ready for this. The safest thing for everyone else and the safest thing for me is to put my hands up and say, I'm not happy here.* His inner dialogue continued: *Am I just copping out? It's not as bad as I might be making it.*

'I think I should maybe go and speak to someone,' suggested Jake, getting straight to the point.

The Boss was very supportive and offered to set up a meeting with the Defence Centre for Mental Health and the psychiatrists. Jake was relieved. In part it was due to having raised his concerns officially, but also because the Boss had responded so positively.

That positive response was appreciated all the more because the loss of an Op 1 crew member – someone who could use weapons – would put more pressure on the Boss at a time when every squadron member was needed.

Jake thought about it overnight, still wondering if he was making a mountain out of a molehill, but also thinking, *Well, I kind of think I need to*. So he went to see the Boss the next day and confirmed he'd like to see someone.

What followed was a multi-step process: see the Medical Officer; get referred to the Occupational Health nurses; then ultimately see the psychiatrist. It would be the psychiatrist who would sign him back as fit to fly with weapons if everything went well.

Two specific factors dominated Jake's thinking. The first was a recurring sensation: *I just can't take anyone's life*. The second was a scenario that he had witnessed a couple of times in the GCS after returning from deployment. The scenario was straightforward and involved making a choice, the kind of choice that aircrew have been making since the First World War. It went like this. There are two individuals walking separately along a road in an IS-controlled area. They are two jihadists who have been positively identified as such, and whose actions make them liable to attack: to being killed. Yet there's nothing to differentiate between the two. And because of the distance between the two, it is only possible to kill one or the other from a Reaper with a single bomb or missile. Even if a follow-up strike was carried out immediately after the first, there is a chance the second figure would have escaped.

Jake's thought process went as follows. *Say they're both twenty-something years old, which many of them probably are. What right do I have to decide which one of you is going to live for another sixty years, have children and grandchildren, and which one of you is going to have your life ended right now. Do I have the right to make that decision? Because effectively, in that operational environment, the decision isn't being made for me.*

It's not that one of the men has a heavy machine gun and one has a light machine gun, and therefore Jake is going to target the greater, immediate threat. Nor can he try and select the greater potential threat, the one who might be involved in the greatest number of killings. But one must die.

Many soldiers have refused to fire upon an enemy for one reason or another. George Orwell famously refused to shoot an enemy soldier during the Spanish Civil War, even though the Law of War allowed it and his orders demanded it.[27] He was not a pacifist but still refused to pull a trigger at one point, with an enemy soldier in his sights.

Jake was very open with everyone around him, including the chaplain, as he wrestled with the idea of playing god – 'with a small "g",' as he put it. The increased frequency of taking shots in this new theatre of operations magnified his self-questioning. However, he continued flying – at least the 'stunt flying' part – over the coming weeks and months, supporting the operation but he was still either unable or unwilling to shoot. He also found it difficult to be around other operators whom he thought were too keen and too proud of a shot. But over that period he did come to recognise his own capacity to 'overthink matters'.

The other thing that Jake gained over that period, as he carried out surveillance and reconnaissance missions against IS, was first-hand observation of just what their fighters were capable of doing to their enemies on the ground or to people whose communities they wanted to control. In his weighing up of the moral balance between lesser evils and greater evils, it surprised him – 'staggered' him, even – to find that the idea of

27 Michael Walzer, *Just and Unjust Wars* (New York: Basic Books, 2015) p. 140-1.

using weapons against this particular enemy was becoming more acceptable to him. He would never be keen on firing weapons but he reached a point of inner acceptance.

The psychiatrist authorised his return to full flying duties. It was up to Jake now.

He went back to see the Boss, around three months after he asked to be downgraded to Limited Combat Ready and Op 2 flying status. 'I've been signed back on. If you are content, I am good to declare myself Op 1 again.'

Minutes later he was back in the GCS and almost immediately was called upon to take his first shot since the fireball in Afghanistan. Helpfully, in terms of easing back in to that part of the job, it was against a piece of unmanned IS military equipment. Despite Jake's willingness to shoot, there was still no new-found enthusiasm for it and he was glad that there was nobody nearby. His next couple of shots were similar: hitting objects rather than people.

Inevitably, after a few weeks, the moment came when Jake was required to fire a missile that was about to result in the death of an IS fighter. The whole process ran smoothly in the build up to the shot, and, before he knew it, the missile was in the air and he was guiding it towards its human target. Despite the focus required to keep the crosshairs on the man he was about to kill, a fleeting, silent thought went through his head: *I'm sorry*. He could see the humanity even in an odious enemy 4,000 miles away. At the same time, he did not consider shifting the crosshairs to an empty piece of land.

Whatever the human connection that prompted his mental apology, it did not change his operational decision. Over the preceding months he had watched IS jihadists carry out too many heinous acts, from firing weapons indiscriminately – and unseeing – around corners, to murder.

Another factor that influenced him was observing a strike on the 'two men walking down the road' scenario. The crew hit the first of the fighters with a missile, and his companion continued to walk past. Jake could not see any sign of the man looking across, no sense of him running or hiding. No sense of 'I've been spared here. I've been given a chance – I must seize it and get away from this place.' No sense of wanting to prolong his own life. No visible reaction. Just an acceptance of what he had just witnessed, then continuing about his business. And so, for Jake, the 'choosing' somehow became more practical, more manageable: part of the responsibility that he chose to retain rather than walk away from.

The advent of the Reaper – of remotely piloted aircraft – has changed some aspects of war beyond what could have been imagined in previous generations. A Lancaster bomb-aimer in the Second World War would not have had to wrestle with the dilemma of, 'Do I kill this individual or that individual, both of whom I can see in great detail?' The challenge then was whether they could hit this town or that town.

The choosing, and living with the choices, is also part of what makes RAF Reaper operations the most human of activities. It is far from public misrepresentations of 'drone' use as some kind of autonomous, robotic, dehumanised way of war. And in the time it has taken to read this chapter, someone, somewhere, may well have had such a choice to make.

In his writings, George Orwell provided many insights into what humans are capable of doing to each other. All the more intriguing, then, that when he wrote about his own experiences of fighting he recalled what he chose *not* to do to an enemy soldier. He fought in an anarchist battalion against Franco's Nationalists in the Spanish Civil War. He spotted a man who jumped out of his trench a hundred yards away and ran in full

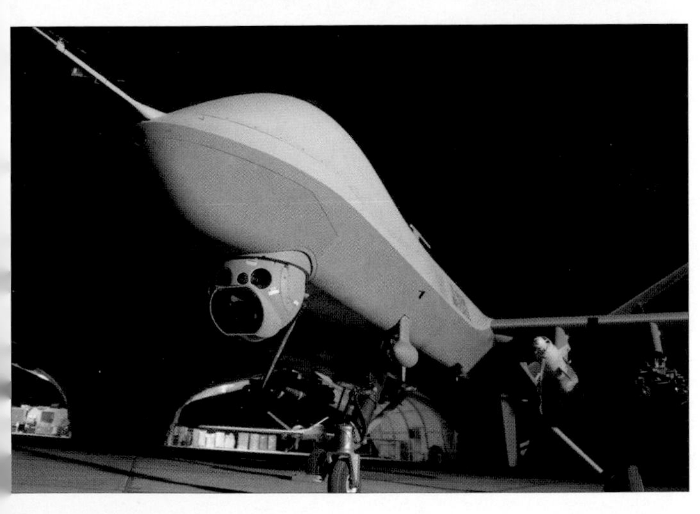

Top: Construction of the RAF 39 Squadron building at Creech US Air Force Base, Nevada, in 2007. *(Photo © Andrew Jeffrey)*

Middle: A 39 Squadron MQ-9 Reaper outside hangars at Creech Air Force Base. *(Photo © Crown copyright 2018)*

Left: An RAF Reaper at night, showing the sensor pod. *(Photo © Crown copyright 2018)*

Above: 'The Squadron Boss . . . personally stuck the RAF markings onto Reaper ZZ200' – the first RAF Reaper to operate out of Creech, October 2007. *(Photo © Andrew Jeffrey)*

Below: A rear view of a Reaper taxiing, showing the 'pusher' turboprop and underwing stores – in this case, four AGM-114 Hellfire missiles. *(Photo © Crown copyright 2018)*

Above: 39 Squadron Reaper pilot and sensor operator at work, thousands of miles from the aircraft they are controlling. *(Photo © Andrew Jeffrey)*

Below: A sensor operator at his controls. The SO's tasks include controlling the various cameras and other sensing equipment, and laser-guiding missiles and bombs onto targets. *(Photo © Crown copyright 2018)*

Top: Its mission ended, an RAF Reaper approaches the military airfield at Kandahar, Afghanistan. *(Photo © Crown copyright 2018)*

Middle: Armed and ready: a night-time shot of a Reaper at Kandahar. *(Photo © Crown copyright 2018)*

Below: Carrying bombs and Hellfire missiles, and controlled by its Launch and Recovery pilot, a Reaper on approach to landing at Kandahar. *(Photo © Crown copyright 2018)*

Above: An MQ-9 Reaper overhead, showing its full complement of underwing stores.

(Photo © Crown copyright 2018)

Below: Reaper at sunset.

(Photo © Crown copyright 2018)

Above: A close-up of a taxiing Reaper, showing the weapons mounted on pylons beneath the wings. These are a total of four AGM-114 Hellfire missiles with, inboard of them, two Paveway 500lb laser-guided bombs. *(Photo © Crown copyright 2018)*

Below: A Reaper and ground crew preparing for a training flight at Creech Air Force Base, Nevada. *(Photo © Crown copyright 2018)*

Above left: Signing the pre-flight documentation prior to a Reaper mission.

(Photo © Andrew Jeffrey)

Above right: RAF Remotely Piloted Aircraft System (RPAS) 'wings' for specialist Reaper pilots, which differ from the traditional RAF pilot badge by having blue, instead of gold, laurel leaves.

Below: A fully armed Reaper taxiing for take-off for a mission in Helmand Province, Afghanistan.

(Photo © Crown copyright 2018)

Above: The author, Peter Lee, with an MQ-9 Reaper at RAF 39 Squadron, Creech US Air Force Base, Nevada, 2018. *(Photo: © James Robertson)*

Below: Corporal Matthew Richard, US Marine Corps, KIA Helmand Province, Afghanistan, 9 June 2011. His story is told in the Epilogue to this book. *(Photo © Joel Anthony)*

view of Orwell and his colleagues. They assumed he was some kind of messenger. Orwell wrote:

> He was half dressed and was holding up his trousers with both hands as he ran. I refrained from shooting him… I did not shoot partly because of that detail about the trousers. I had come here to shoot at 'Fascists', but a man who is holding up his trousers isn't a 'Fascist,' he is visibly a fellow-creature, similar to yourself, and you don't feel like shooting him.[28]

Orwell did not 'feel like shooting' the man in whom he fleetingly glimpsed an element of common humanity. So he didn't. Michael Walzer writes about several such examples from different wars, where legally and morally entitled combatants – soldiers – refused to fire on an enemy.[29] In those instances, the common element that discouraged them from firing was seeing a fellow human being and not some dehumanised monster.

The view on the screen that a Reaper crew member gets of an enemy fighter will be similar to what Orwell could see at 100yds with the naked eye. Except Reaper crews can watch a particular individual for hours, days or weeks before being required to target and kill that person. They do not just get a glimpse of the humanity in their enemy – they see it in great detail. As one Reaper pilot described it to me:

> The distance is irrelevant and, if not, it is a positive thing. It is often argued that because we're distant from the target, unlike a fast jet, the process is dehumanised or that we are separated from the process. The boundaries of time and

28 Michael Walzer, *Just and Unjust Wars* (New York: Basic Books, 2015) p. 140-1.

29 Ibid, p. 138-44.

space, however, are overcome by the technology; we're not distant from the target. The target is in my eyes all of the time; I can see everything he does.

Another Reaper pilot went even further in describing what it was like to engage with that human dimension of war:

> Both conventional aircraft and the Reaper work under the same rules of engagement during a weapon strike. When I am in the GCS, I *am* in Afghanistan and I *believe* that in certain circumstances we have the ability to have more situational awareness and involvement in a target than a conventional aircraft. For example, we may watch 'Target A' for weeks, building up a pattern of life for the individual: know exactly what time he eats his meals; drives to the Mosque; or uses the ablutions – outdoors of course! This is all-important for the guys on the ground. However, what we also see is the individual interacting with his family – playing with his kids and helping his wife around the compound. When a strike goes in, we stay on station and see the reactions of the wife and kids when the body is brought to them. You see someone fall to the floor and sob so hard their body is convulsing. A conventional aircraft often doesn't have the endurance [in the air] to witness this. I believe [that] although we are sat in the UK conducting strikes from afar, it is these situations that bring humility.

When a Reaper crew prepares to fire a weapon that will kill someone, they often do so with a detailed and intimate knowledge of the life that is about to be ended. They have more awareness of the basic humanity of their enemy than almost any other combatants have had in the history of warfare. And that is the background to the dilemmas that Jake wrestled with: 'Who am I to take life?'

He also said, 'If I think back to the shot I took before I went away, it was against an armed specialist team moving towards friendly forces. In your mind, that's a reasonably straightforward one to justify: I'm protecting life by taking life.' As for so many Reaper crew members, the 'protecting' element is central to how he thinks about his involvement with the Reaper. He then added, 'Now [in Syria and Iraq], there isn't that direct threat to coalition troops on the ground. I don't have that same protective instinct. I have it from the more generic sense of removing a broader threat from a battle space rather than saving an individual life by taking life.'

Those fine distinctions clearly shape not only how he thinks logically about using weapons, but also how he feels about it and the emotions they prompt. And those thoughts and emotions have changed over time. As he explains…

> There is a last thing that I think is worth bringing up. At the start, weapons events were very rare. I would get that surge in the pit of my stomach when I watched. Not just when I was taking part in a shot, but when I was watching one, or if I was in another GCS and looking across at another video feed on the screen. I would be transfixed by it and I'd be thinking: *This is a terrifying, terrible act.*
>
> Now, when I'm the Authoriser and controlling the missions from the Operations Room, or even when I'm in one of the other boxes, when someone says they're about to strike, I look up. I see the explosion, see people walking around, moving around with what they're doing. I note the time if I need to for the log and I carry on my day-to-day work.

> In that first period of time on the Reaper, I could recall every shot that I took – and I only took a couple – and most of the other shots that I saw. Now [in the fight against IS], the frequency of shots has gone up. That has led to a natural blurring of the memory. I have probably shot much less than the average person on the squadron now. But even I cannot, 100 per cent, if you asked me now, reel off the shots I've taken.
>
> There have been a couple of occasions where the number of individuals was high under a given shot. And I'm pretty aware of what, to use the awful term, the 'body count' is. But for my own mind, to not be fully aware of the lives that I've taken, or the lives that I've changed, does make me slightly disgusted with myself.
>
> Now I completely understand that as the frequency of using weapons goes up, that will always happen. I always prided myself on having a very good memory with these things, but I do wonder, have I become less moral?
>
> I hate the fact that I've killed people. I hate the fact that I've taken life in a very calculating manner. But I also hate the fact that I seem to be able to live with it now. Because, to my mind, whether you fly a fast jet or whether you fly the Reaper, and you do take life, you shouldn't really be able to just get on with things without thinking about it. To me you've crossed the line there. But you can. It is an inescapable element of being in the military.

Everyone who flies the Reaper will make hard choices about life and death, and live with those choices. They will have a close-up view of what they do and they will retain the mental images of the aftermath. So here is a choice for you, the reader, to wrestle with. It is a hypothetical choice based on real-life events.

There have been numerous reports from Iraq and Syria of men being sentenced to death for being gay. IS has released a number of videos showing gay men being executed in Raqqa and Mosul. They were thrown from a tall building and then stoned to death afterwards if the fall did not kill them immediately.

Picture yourself in a Reaper GCS, your screen filled with the image of a holding cell or cage containing several men, who have all been detained for being gay. Yesterday you watched as two armed IS fighters from the local militia arrived and selected one of the men from the cell. They dragged him to a three-story building a couple of hundred yards away, then threw him off the top to his death below. For being gay.

Today, you watch the same build-up as another execution is about to happen. Except you have a Hellfire missile at your fingertips and you have the option of killing the two IS fighters – crucially, without harming anyone else – before they select the next gay man to be killed. They have been positively identified and authorisation to strike is in place.

What do you choose? To kill the fighters and live with the close-up images of your actions? Or leave the fighters alone and watch – in detail – as they kill a man whose only 'crime' is to be gay? Plus, the fighters are too far apart to hit with one missile.

Choose.

CHAPTER 9

HAPPY BOXING DAY

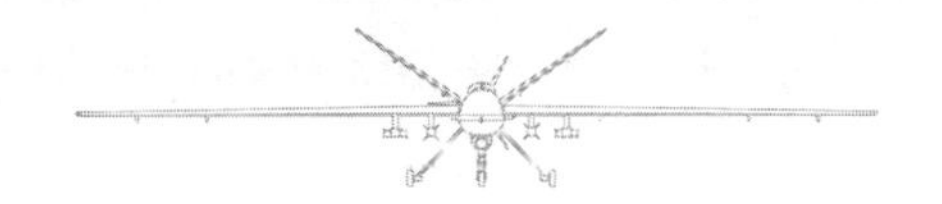

'I KNEW INSTANTLY THAT IT HAD NOT BEEN A GOOD NIGHT. YOU COULD JUST TELL HE WAS VACANT.'

PENNY, A REAPER PILOT'S WIFE

The day was as normal as any day can be when you stand guard at the gates of hell. The tall figure was clearly in charge as he peered from the shadows of the guardhouse to watch the young soldiers – boys really – manning the checkpoint at the far side of the bridge. Every time someone on foot or in a car approached he shifted position slightly and so did they. Getting ready to respond to the IS threat that was never far away. If there were Iraqi regular soldiers among them, they were mixed with either militia fighters or novices who did not even have a standard uniform.

The three at the checkpoint were part of a deployment of around two dozen men. The rest were either sleeping, cleaning weapons, smoking or loitering near the cookhouse. The road they were occupying came right through IS-held territory and across the bridge in front of them. If there was such a thing as a border then they were guarding it. For a FOB it was as 'forward' as it is possible to get. Most of the nearby buildings in what was once a trading town were now abandoned.

A tense, uneasy calm had held for a few weeks. There was no clear intelligence on whether IS fighters had halted their advance or whether they were waiting and preparing for another push. Fight or run, that would be the choice at the guardhouse if the attack came. Anyone who got captured would likely be tortured and killed.

Many soldiers on the front line against IS kept one bullet back at all times in case they were overrun by the jihadists. A split-second decision would have to be made: suicide or risk being brutally murdered in an orange suit on YouTube. The favoured IS approach for maximum social media impact was an extended line of terrified Iraqi soldiers having their heads hacked off. It wasn't a quick process.

As they stood guard waiting for the inevitable attack, were they even aware that it was Christmas Day?

For Gav, Christmas Day in 2014 was like any other. Or at least like any other Christmas Day when you were going out to look for someone to potentially kill. Not just anyone, obviously. No, the guys they would be looking for today were not going to let a little matter like infidel Christmas celebrations get in the way of expanding their caliphate in Iraq.

Gav could feel his mental 'business' compartment opening and expanding, and his 'family' compartment shrinking and closing at the same time. He had gone through this routine so often that he could actually sense the change as it was happening. He didn't know how he did it or how it started. All he knew was that there was not enough space in his head for both of those compartments to be fully expanded and open at the same time.

As he sat waiting for the pre-flight briefing to start, thoughts of Penny and the kids drifted away. He was happy to stop thinking about how they were when he left them less than an hour ago: sprawled out and half-asleep watching *Chitty Chitty Bang Bang*.

It takes careful planning to perfect the nonchalant getaway from home. The kit bag, coat, shoes, packed dinner – turkey leftovers – and car keys all get deposited next to the back door in advance. Gav pretends not to be getting ready to go out; Penny and the other visiting family members pretend that they don't notice. If nobody says anything then reality can be delayed for a few more minutes. It was almost 11pm when he paused as he passed the living room door, catching Penny's eye. A shared smile – he would see them all tomorrow.

'Merry Christmas, everyone!' The Auth jolted him from his reverie and back into the Briefing Room, the 'family' compartment slamming shut.

The customary response to the standard Yuletide greeting is to politely echo, 'Merry Christmas.' On this occasion, responses ranged from an ironic 'Yeah, Merry Christmas to you too,' to mild boos, one threat of violence, a couple of whispered invitations to depart in a sexual manner and the throwing of a paper aeroplane.

It was a good effort on the part of the Auth to raise morale. Somewhere in the mists of time – probably to relieve boredom or pressure – duty officers started to introduce the odd light-hearted note into the pre-flight brief. Not every day, just often enough to stay within the tolerance of the CO and the Squadron Executives.

On this Christmas Day, though, there would be no chance of the Auth going beyond what the Boss considered acceptable briefing banter, because the Boss was the Authorising Officer.

He had given another Auth the day off. Across the Atlantic, the Boss of 39 Squadron had done the same thing.

One element in a good briefing has always been to correctly read the mood of the room. And it was clear no one wanted to be there, partly because the main task was not the dynamic and adrenaline-inducing taking the fight to IS. Instead, it would be a continuation of the 'watch and report' activities of the past couple of days. Gav and his crew were given an area where IS fighters had recently been making gains against Iraqi forces on the ground. Humiliatingly for Iraq and its allies, some of those advances were being achieved using high-quality American weapons that the jihadists had acquired when they overran previous Iraqi Army positions.

As the briefing came to an end Gav ignored what was going on around him as the relief crew and others started to file out of the room. He was double-checking his notes on the task that lay ahead: hours of scanning an IS-controlled area looking for enemy activity. Looking for 'badness'. He also made a note to check on the airframe he would be flying.

'See you for the out-brief in ten minutes,' he said, making his way to the door. As he walked to the crew room to make a 'brew' he overheard several snippets of conversations and noticed that, unusually, nobody was talking about their families or what they were doing before they came in. It unsettled him for some reason.

A trio (Gav, Dunc and Marty) arrived at the Operations Desk for the out-brief before walking out to the GCS. The Boss-Auth was determined to maintain an air of Christmas bonhomie but the few ratty strands of tinsel and the sad, almost bare mini-Christmas tree behind the desk – surely rejects from someone's attic clearance – lent a depressing air.

'Gentlemen,' interrupted the Boss-Auth, 'is everyone fit to fly?'

The three parroted, 'Affirmative'. The same rules on alcohol that apply to the pilots and crew of civilian airliners and conventional military aircraft also apply to Reaper operators. Christmas lunch had been a dry affair for everyone.

The rest of the briefing followed a long-established pattern that everyone from a flight cadet in training to a senior combat pilot would recognise.

'NOTAMs' [the Notice to Airmen warn aircrew of mandatory areas to avoid, and potential risks in the area where they are planning to fly]. A stream of hieroglyphs was distilled down into two simple warnings. First, expect other aircraft to be flying in and around your designated area of operations. Second, there is a lot of stuff going on over Northern Iraq. Gav wondered if Second World War briefings predicted the possibility of German aircraft over southern England during the Battle of Britain.

SPINS [Special Instructions] followed, attached to the air tasking order. Nothing that they need to worry about today. Aircraft flown by coalition partners would be staying out of their airspace as much as possible. A Reaper can be almost invisible to a fast jet in certain circumstances and the Top Gun boys and girls like to know when the plodding remotely piloted aircraft are around.

'No concerns over the airframe today – you're all good to go.' The Boss-Auth was confirming for Gav what he could see for himself on the sign-out sheet. Every minor fault with the system, from the GCS to the airframe, was captured on the sign-out sheet. Once he signed the sheet, Gav would be responsible for the aircraft. The process always reminded him of hiring a car. Identifying and signing for all of the dents and dings before you drive the hire car away, then telling the company about any faults when you bring it back. At least the theory worked with the Reaper. Who in their right mind would own up to a fault or a blemish with a hire car if you might get stuck paying for the repair? However, when weapons are involved, attention to detail is crucial.

The stay-out-of-jail check came next: RoE and Law of Armed Conflict. Dunc and Marty were as interested in this bit as Gav: they would all be up on the same charges if they killed the wrong person in the wrong place at the wrong time.

Gav already had eighteen missile strikes to his name. With each strike he became more, not less, aware of the consequences. During the flight, though approval of a particular strike would come from the RCH at the CAOC via the ground controller, Gav still had to pay attention. If he did not like what he was seeing as he was getting ready to fire – for any reason – it was up to him to not pull the trigger.

He ran through his own checklist, confirming it to the Boss-Auth. This last-minute check confirmed he had noted down all the correct information: an outline of the day's op; the radio frequencies he would be using; the weather that he was expecting (occasional limited visibility); and that he would approach either static or moving targets, as someone once said of porcupines mating, very carefully.

'Have a good one.' The Boss-Auth had dropped the Christmas theme and gave them his usual farewell as they disappeared towards the hangar and their appointed GCS. They took their seats and started to get comfortable, adjusting headset volumes and setting up their own workstations. At the back of the box, Marty's workstation and all its intelligence feeds looked more like the desk of a city trader than the cockpit of an aircraft. The live video feed from the designated Reaper dominated his view. Gav and Dunc had the same picture at their pilot and sensor stations: a small hamlet being watched by their colleagues on 39 Squadron at Creech.

Then there were more checks to set up the GCS. As soon as Dunc was settled in the right-hand SO seat he got out the checklist. After a nod from Gav he started reading at breakneck speed. Gav's hands were a blur as he responded to the instructions. At the same time, he gave Dunc a quick-fire series of replies: multiple 'confirmed', 'green', 'good' and the occasional 'happy'. Then all three started translating handwritten notes from the pre-flight briefing into a series of rapidly typed entries using the well-worn QWERTY keyboards in front of them.

'Handover checks.' Gav moved on to the next task. And so they continued. Prevailing wind conditions – over Iraq – were hastily scribbled on an old-fashioned office whiteboard on the wall next to his head. Dunc and Marty were writing up their own encoded reminders.

Fuel and weapon status. It was always at this point when Gav checked the weapon racks on the aircraft and noted the current payload that reality – or what stands for reality in a cockpit thousands of miles from its aircraft – hit home. A brief telephone conversation confirmed that four Hellfire missiles and one 500lb GBU bomb were about to be at his disposal. A small flicker of adrenaline. Was he excited at the prospect of using them? Gav wasn't sure but the thought preyed on his mind occasionally.

Communication and satellite uplink switching completed the picture on Gav's main screen. Once he was in full 'shadow mode', he could see what the pilot in Creech could see. The only difference was that his controls for flying the aircraft didn't yet make it change direction. Dunc and Marty were also fully in sync with their counterparts several time zones away.

Then, finally, Gav spoke the crucial words that have been uttered for more than a century as command of an aircraft changed hands: 'I have control.' Another blip of adrenaline.

Dunc echoed 'I have control' from the SO seat as he took over the camera pod.

No trundling down the runway, slowly climbing and transiting to the operating area. They were instantly in full surveillance mode in a far-away war zone with a full weapons payload. There was no easing into the job during an airborne handover. It was the mental equivalent of a sprinter in a relay team being handed a baton while exerting maximum effort.

The first hour passed slowly as they broke their search area into small sectors and circled above each in turn, watching buildings. There was nothing unusual down below. But nothing was normal either. In the dead of night some of the small villages and isolated houses looked like deserted film sets. The inhabitants had obviously fled when IS took control of the area.

Gav briefly wondered where the people were. Dead? In refugee camps? Somewhere in Syria? Packed in an overcrowded boat attempting a lethal winter crossing of the Mediterranean to Europe? The crew spotted the tall Iraqi commander and his men, and spent a few minutes checking them out.

'Routine checkpoint,' said Marty, 'Iraqi Army FOB.' He called the SMIC in the Ops Room upstairs to double-check the type of weapons the soldiers were carrying. From a separate intelligence computer chatroom he confirmed that the checkpoint was still being manned by friendlies.

'All quiet,' said Dunc as Gav brought the aircraft round to a southerly heading. Every car they spotted was scrutinised but there was no sign of any IS 'technicals', the fast-moving, highly prized pickup trucks with heavy machine guns or artillery pieces on the back.

Marty broke the brief calm: 'There's badness here. I can smell it.'

Gav and Dunc smiled. Marty was always smelling 'badness'.

'There's a reason you're sitting back there,' replied Dunc. 'So we don't have to smell your "badness".'

As thoughts started to turn to the arrival of the relief crew and a table full of mince pies in the crew room, Marty jabbed at his screen.

'That's interesting. Look there!'

'Where? Any more clues?' asked Dunc, who could not have seen Marty's screen if he tried.

'Sorry. Bottom left of the screen – just came into view. Dunc, can you zoom in?'

All three of them watched intently as the image increasingly filled the screen.

'It looks like an M113 armoured personnel carrier,' said Marty.

They all had the same thought at the same time: What is *that* doing *there*, heading *out* of Daesh territory?

Gav contacted the JTAC to report a suspicious sighting. Armoured personnel vehicles like this almost never travelled alone. Where were the support vehicles? Where was the rest of the convoy?

Within seconds came a directive over the radio: 'Keep eyes on the target. Ground forces have been looking for a stolen Iraqi armoured personnel vehicle in this area since yesterday.'

Gav requested approval for a strike and commenced his 9-line checks with the JTAC.

'Village ahead,' announced Marty. As soon as the vehicle got into the village there would either be people nearby or inhabited buildings. He reckoned that it really was a vehicle that had been captured by IS. He was working his radio and intelligence chatrooms to positively identify the vehicle as an IS weapon.

Meanwhile, Gav worked on strike permissions and Dunc kept his camera trained on the vehicle. In response to a request, Marty started digitally rewinding the footage and checking the detail frame by frame.

'Confirmed hostile, 9-line approved,' said the JTAC. Gav's adrenaline surged, the only outward sign being a slight tapping of his left heel on the floor under the desk. He acknowledged his permissions and went straight into his pre-strike checks. As always, the UK RCH was crucial in providing legal authority – his stay-out-of-jail card. As the M113 trundled up the road it did not seem to be in a particular hurry. Nobody was manning the heavy machine gun on the top.

Gav started to bring the aircraft around to get an optimum strike angle for a Hellfire: the armour piercing missile was originally designed for jobs like this.

'Vehicle entering village,' called Dunc. 'No strike. It looks like it is slowing down.' He kept his camera focused with one hand and the vehicle in the cross-hairs with the other. Then the vehicle tracked off the main road and pulled up at some kind of car shelter attached to a building. The building seemed too big to be a normal house. Three men climbed out and started covering the vehicle with some kind of tarpaulin. The whole crew was well aware that if they had not watched it they would never have known the vehicle was there, apart from the massive heat signature from its still-hot engine. What else did they see but not see?

The JTAC came over the radio and told them what they knew already: 'No strike. Confirm no strike.'

'No strike confirmed,' responded Gav. The silent staccato beat of his left heel slowed slightly.

The minutes dragged as the Reaper flew its observation pattern overhead. People came and went from the main building, ducking in from smaller dwellings nearby. The view from 18,000ft was unobstructed by cloud, and other aircraft were being kept away from them. Even from that height, though, it was clear that the movements were not natural: too rapid for the normal tempo of rural Iraqi life. Also, the fighters stayed as close to the walls as possible and some of those coming and going carried weapons. IS leaders did not tolerate opposition within their territory so the weapons probably belonged to their fighters.

'Can we assume that these are not social workers?' asked Gav.

'That depends,' replied Dunc. 'Yes, if we are over Iraq. No, if we are over Glasgow.' Marty ignored the barb that was aimed at him. He did smile though. Gav's leg had stopped drumming its manic beat.

After what seemed like an eternity, but which was probably fifteen minutes or so, three men came out, uncovered the vehicle, unslung their rifles, clambered in and headed north. As soon as the vehicle cleared the village – plus the distance of a generous blast radius – strike approval would kick in again and another significant IS weapon would be permanently disabled.

'Confirm 9-line,' said Gav to the JTAC, 'target has cleared the village.' The foot tapping resumed at full throttle. Gav's voice betrayed nothing: he could have been ordering pizza. The preternatural radio monotone of RAF aircrew was sacrosanct. He would have sounded *more* stressed ordering pizza.

'9-line confirmed,' came the response. 'Cleared hot.'

As he double-checked his missile selection and its blast setting, he checked with the others: 'Everybody happy?'

'Roger. Target locked on,' replied Dunc.

'Roger. Road clear ahead. Safe shift area anywhere to the east of the road.' If anything happened – like the arrival of a civilian or unidentified car – in the last few seconds before impact Dunc would 'drag' his cross hairs and the laser 'pointer' they control to an empty field. The missile would follow and its impact be absorbed by some piece of dirt.

Dunc's hand was tense as he gripped the cross-hair joystick. Whether the Hellfire hit the vehicle or not would be up to him. His target was fairly slow moving, making the shot somewhat easier. He tried to ignore the mental image of the number of people who would be watching this shot: at least a handful in the Ops Room upstairs, and possibly many dozens in the CAOC. *Don't fuck it up*, he told himself.

Gav started his count-down, his trigger finger relaxed. His foot was channelling all of his tension: 'Three … two …'

Then, from nowhere, 'ABORT. ABORT. ABORT.' The JTAC's voice erupted into their headsets. 'Confirm that there is no weapon in the air.'

Gav's finger flicked back off the trigger as though it had received an electric shock. 'Nothing. No weapon in the air,' he responded, just about keeping his voice in check as he visually confirmed from his screen that the four missiles were still in place. He spoke to the other two: 'Did we miss something?'

'Negative.' Dunc sounded adamant.

'Keep eyes on the target.'

'There's some chat coming through on the intelligence,' added Marty.

Gav asked the JTAC why the strike had been aborted. There was a question from the RCH, who had wanted to double-check the PID of the armed personnel carrier as definitely under IS control. 'Marty, is it a friendly?'

'Waiting confirmation.' The confusion had only been going on for fifteen to twenty seconds but the whole dynamic had changed. Uncertainty had been introduced. 'Gav, target is still moving north, eight hundred metres from friendly checkpoint.' Pause. 'Something isn't right.' Marty started counting down the distance – 'seven hundred metres…six hundred…'

As the jihadists' vehicle got to within five hundred yards of the checkpoint, it passed the last point at which they could safely attack it with a Hellfire. At the speed the armoured vehicle was moving, by the time a Hellfire reached it the Iraqi guards would now be in the blast radius.

The JTAC announced over the radio: 'Confirmed hostile. You are cleared hot.'

'Cleared hot,' confirmed Gav. Except they could not fire now without killing or wounding the Iraqi friendlies.

'Shit!' Marty did not shout at anyone in particular. Silence followed. The target vehicle suddenly accelerated towards the checkpoint. The sudden increase in its dirt trail gave away the acceleration. As it lurched forward they watched someone clamber up on top to the 0.50in-calibre machine gun position and start firing at the soldiers ahead. That .50 cal could blast its way through a brick wall.

The silence hung heavily in the GCS as the three crew members made the same assessment: they could still hit the armoured vehicle with a Hellfire but the blast would also kill the nearest soldiers that they were supposed to be protecting.

They could only watch as two of the soldiers at the checkpoint turned and started running across the bridge behind them, away from the IS attackers. The third was down. Shot. They could see the other Iraqi soldiers rushing to take up defensive positions at the north end of the bridge but their small arms fire would not even dent the armoured vehicle racing towards them.

The M113 had tracks instead of wheels and tyres. As it approached the concrete chicane at the checkpoint – giant concrete blocks that forced cars to carefully wind their way past – it did not even slow down. It smashed through the concrete blocks and over the dead body. The handful of soldiers at the far end of the bridge scattered as the juggernaut approached, while reinforcements spilled out of the guardhouse.

Everything seemed to go into slow motion and then the camera screens in the GCS blanked out.

'VBIED.' A giant car bomb. Marty had seen one recently and the effects were fearsome –much more powerful than the blast of a Hellfire missile. As the hot dust cloud from the explosion started to subside the crew could only watch in silence. Most of the guardroom had simply evaporated. Dunc could tell from the thermal imaging that three Iraqi soldiers were on fire, and one of them had almost made it to the other side of the bridge before he collapsed.

'We could have had it,' whispered Gav, pointlessly, as they watched the carnage. The others just nodded silently.

There would be no let-up. The final straw came in the form of a technical that emerged from among the disused buildings only a few hundred feet away. The jihadist on the back blasted the surviving Iraqi soldiers with a Dushka, the DShK 0.50in-calibre Russian heavy machine gun.

'Is there a shot?' asked Gav, knowing full well there wasn't, but desperate to do something to rescue the Iraqis below.

'Negative.' Dunc wanted to evaporate the technical just as badly. 'Surviving friendlies are still in our blast radius, and we don't know who or what is in the nearest buildings.' Whatever way they looked at it there was no way to safely intervene.

It was a massacre. From the IS perspective it was a brilliantly planned and executed massacre. From 18,000ft overhead it was a slaughter. Worse, it was a preventable slaughter, in Gav's view. He called the Boss-Auth, who would be watching in the Ops Room. Unbeknown to Gav and the crew, a steady trickle of other squadron members had wandered in to watch the live video feed.

'Boss, can you find out why clearance for the shot was withdrawn. We would like to know why we are watching friendlies on fire.' The seething lightness of Gav's tone was worrying. Few things hurt in the military like letting down people who are relying upon you. If Gav was going through some five-stages-of-grief type process then he had very rapidly progressed from disbelief and denial, embraced anger and was heading rapidly for bargaining.

On one level he knew, logically, that nothing he could do or say would change the situation. However, on another level existed a desire for some kind of cosmic justice. If he got a satisfactory answer to his question about the time delay – went his reasoning – then he could make better sense of what he was watching. He could live with it more easily.

'The delay was procedural – a double-check on possible collateral damage. It only took a few seconds,' replied the Boss-Auth. Gav was not reassured. *Fifteen bloody seconds.* The Boss-Auth judged that Gav was also not ready to be asked how he would be feeling if it turned out that an error had been made and that Gav had killed friendlies himself by firing upon an Iraqi-manned vehicle. 'The relief crew is coming down now. Grab a brew then I'll come and chat with you.'

'Roger.'

The crew stepped silently from the GCS into the chilly gloom of the hangar. They trudged up the two flights of stairs to the Ops room.

'What do you think the Boss wants?' asked Marty.

'To congratulate us on a job well done looking out for that Iraqi FOB.' Gav was in no mood for small talk. Dunc kept his thoughts to himself.

The crew room was deserted when they walked in. 'I guess we're the lepers of the day,' observed Gav. When something bad happened and nobody knew what to say, sometimes it was just better staying out of the way.

As they were pouring their drinks, the Boss walked in and

sat at the breakfast bar behind them. With his tightly cropped hair, he could pass for a retired welterweight boxer. His flying suit was reasonably new and had all the correct patches on it: XIII Squadron, pilot's wings, callsign, wing commander rank slides. The ancient, worn flying boots were part of a different story, however. They had seen him through thousands of hours of Harrier fast-jet flying. The others joined him at the breakfast bar.

'Tough break guys. I caught everything on the feed.' With those words, his well of sympathy started to run dry.

'We had him, Boss,' stated Gav, his foot tapping again.

'It was just one of those things.'

'Who caused the delay?' Gav hadn't given up on bargaining with history.

'It doesn't matter. They had to double-check and you know it.' Sympathy left town. 'What I need to know is: Are you all fit to fly? Can you go back in and focus?'

They glanced at each other and nodded. It hadn't occurred to any of them that they might not go back in.

'If in doubt, sit it out.'

Gav managed a half-smile: 'That's not going to happen, Boss.'

'Shout if anyone changes their mind. Me and the SMIC will be keeping a close eye on you from upstairs.' With that he got up to leave. 'Before I go, there's the TRiM question. Anyone want to talk about this one?'

No one replied.

After the Boss left, Gav, Dunc and Marty spent just a couple of minutes more discussing what had happened earlier, then started planning what they would do next. Dunc and Marty would finish off the battle damage assessment if the relief crew hadn't finished it. They would need the information to write up an incident report. Gav would find out from the CAOC if they were to stay on the same tasking or if there was anything new. Rarely had a few hours of mind-numbing, routine observation been so appealing.

By the time they returned to their seats in the GCS, the IS pick-up truck and its Dushka heavy machine gun had gone. Another air asset had been given the task of following its departure.

'Remind me to follow up on that later,' said Gav, 'I'd like to know what happens to it.'

'I think I can tell you right now what will happen to it,' Dunc replied, 'they'll be lucky to survive the night.'

'I don't think they'll care,' added Marty.

Nobody answered. How do you try to understand an enemy that doesn't mind dying?

As Gav flew the final orbit around the attack area, from the infrared images on the screen they could see the heat draining from the bodies of the dead Iraqi soldiers. In their midst, a solid white lump revealed the latent heat in the self-destructed armoured personnel vehicle. During the body count they noticed that one of the cooling corpses was noticeably longer than the others. The commander was now a statistic in a column. His lasting memorial would be an anonymous place in an archived segment of video footage and in the mental images of those who recorded it, unable to protect him. Those watchers would never know his name.

Then they received their new tasking. Another IS technical had been reported about ten minutes flying time away. Gav brought the Reaper round to a south-westerly heading, leaning ever-so-slightly to his left as he sustained the long, banked turn. Almost every Reaper pilot does it to some extent – the more extensive the previous flying experience, the greater the brain's expectation that the body will move in tandem with the airframe.

Once they had sight of the new target everyone went into their usual routine. Marty set about confirming PID of the vehicle. 'Shall I ask them to really, really confirm it this time?' He hadn't said much at all in response to what had happened earlier. When it came to it there had been a brief delay for a double-check in the intelligence trail. It hit him harder; almost a sense that 'his' people – intelligence staff – had let them down. Like Gav, he also was not yet able to wrestle with the fact that the delay could have prevented them from killing friendlies. Sometimes in war, shit is not what happens but what doesn't happen.

'I think they'll really, really confirm it without being asked,' added Dunc. They all knew that Marty would never say such a thing over live comms – he just needed to get it off his chest.

'Request 9-line.' The carpet under Gav's heel was taking a gentle, high-tempo pounding from his heel. The approval process ran smoothly, authorisation for the strike was granted.

The pick-up truck had been moving under cover of darkness, its driver not realising that it was lit up like a searchlight on the Reaper's infrared displays.

They went through all the usual checks – direction, speed, weapons, identifying the safe 'shift' zone – when the call came through again from the JTAC: 'Cleared hot.'

A greater-than-usual spike of adrenaline all round.

'No shot,' came the call from Marty. 'Houses ahead. Other vehicles on road.' Gav confirmed the situation with JTAC.

The pick-up pulled in amongst a small group of houses. Even in the dark the IS fighters covered it over with some kind of tarpaulin. Its engine heat served like a pointer for the infrared. They would fly a holding pattern. Gradually the adrenaline levels in the GCS subsided back to somewhere approaching normal.

Before the end of the shift, Gav was 'cleared hot' twice more. Twice more there was no shot. Twice more his finger came off the trigger. Usually when that happened the crew speculated about what an enemy vehicle and its weapons might get up to if it escaped. There was no speculation today – they would never forget what their technical did at the Iraqi FOB.

When the shift finally ended, they went through the in-brief process at the Ops Desk. As the Boss-Auth read through the checks, the various responses came thick and fast. Then for a second time he mentioned it. 'Any TRiM-worthy events?' He paused. No one answered. 'That's a "Yes". I will set it up [psychological support] for a couple of days' time.' And the final question: 'Are you fit to drive home?' The Boss-Auth took a good look at them as he asked.

Three affirmatives. None of them had too far to go and the adrenaline from the final non-shot would last long enough to get them home. The rising morning sun would help as well.

Gav waved the others off as he climbed into his car.

'See you in twelve hours!' Marty had a certain drollery that would not go amiss in an undertaker.

Gav wound his way home through the lanes of Lincolnshire. His only company on the route added up to four cars, two tractors, a dead badger by the side of the road and his own thoughts. Pulling up outside his house, he could see that everyone was up and about, including his in-laws. No doubt the kids had seen to that. He had never been so happy to arrive at a door he did not want to enter.

As he stepped indoors he was regaled by a disjointed chorus led by two highly excitable boys: 'HAPPY BOXING DAY!'

He managed a wan smile but Penny had already read his face. 'Bad one?' she whispered when she got close enough. It managed to be both a question and a statement. Gav just nodded. She handed him a breakfast beer and told him to go upstairs and run a bath. She would come up and speak to him in a while.

'Happy Boxing Day,' he toasted everyone as he left the room

for some minutes of solitude. He wasn't ready to embrace all the happiness yet. *Fifteen bloody seconds.*

When Gav recounted his story to me, I was struck by the seemingly innocuous detail that his wife Penny had taken just one look at him and known it had been a bad night. I mentioned that I would be very interested to hear Penny's perspective on what happened when he walked in that morning. Not that I disbelieved him, more that I wanted a second take because I was slightly surprised by the extent of his self-exposure.

All the years I had previously spent as a military chaplain had taught me a few things about military personnel. Years of training and social conditioning allied with – hopefully – a strong sense of duty and personal integrity usually resulted in a great deal of emphasis on honesty amongst friends and colleagues. (Usually. The military gets its share of nutters and crooks, just like every other part of society.) This should not be a surprise: if soldiers, sailors and airmen train together, kill together and risk their lives for one another then it is likely that honesty is part of that mix if social order is to be maintained. Generally speaking, however, I never found their externalised honesty – words and behaviour – to be matched by individuals' honesty with themselves.

A couple of weeks after I interviewed Gav, I sat down with his wife Penny at their home to get her version of events. By way of background, she tells me how they had met at university.

'The thing is, I always said I would never marry a military man, purely because of the amount of time they spend away. Definitely not Navy and certainly not Army.' Which leaves the RAF, presumably. She chuckles under her breath and murmurs 'Oh God,' in some form of admission.

I can't tell whether she is laughing at her poor judgement in not following her youthful instincts or the fact that she has just admitted that poor judgement to a complete stranger. This is a surprisingly common aspect of conversation with many military spouses and partners. No matter how much they love and support their other halves – or not – there seems to be a common need to explain, justify or apologise for finding themselves married to the military. I wonder if my wife Lorna did the same.

Penny has supported him through his whole air force career to date, though his time on the Reaper Force has not exactly turned out as anticipated. Their thinking was that by volunteering for a posting to XIII Squadron at RAF Waddington, the family would see him more often – every day – than if he worked elsewhere or in a different role.

'How is that working out for you?' I asked, deploying my razor-sharp interrogation skills.

She laughs, again. I am beginning to wonder if she is laughing at me – and I know I'm not that funny – or at her own naiveté. A shift pattern of six days on and three days off doesn't sound too extreme, though it means a whole weekend off with the kids comes around about once a month. Also, depending on the timing of the shifts – she mentions a period of him leaving home at 5.30am and getting back at 9.30pm – he can go for a week or so without seeing the children awake. Again, perhaps not unusual when compared to traders in the City of London.

The big difference seems to be that in the current intensive operations against IS, he is either killing people, looking for people to kill, reflecting on whether they could have killed someone more effectively or gathering information for someone else to pull a trigger. It adds a whole new dimension to the managerial concept of Performance Development Review. He is also tired. A lot.

'He'll say, "I'm fine, I'm fine, I'm fine" until I look at him the right way and he'll say, "All right then."' And she laughs, *again*.

If she didn't laugh I wonder to myself if she would cry instead. I don't ask. She continues, 'I know he enjoys it. From a flying point of view he enjoys the actual job, and I think emotionally he copes very well with it. Even when he does his silly [long] days, he'll come home and we'll be sitting in the evening watching telly and he'll be writing in his note pad – things he's got to do the next day, or instructional things. I don't think he stops thinking about it *at all.*' Italics can't do justice to the stress she put on those words.

Intriguingly, one of Penny's observations stands out as counter-intuitive, when she refers to him deploying to the Middle-East for four months the previous year as part of the Reaper LRE.

'I think it did him a lot of good going on the LRE, and I was looking forward to him going. Not to get rid of him, but I just knew he needed a break.'

For many service personnel and aircrew on other types of aircraft, deployment to an operational theatre overseas is often the high point of the two to three year operational stress cycle. But what does it say about being on a Reaper squadron that going away to an operational theatre for four months is considered a rest?

Instead of pursuing that point, I ask her what she remembers of the Boxing Day morning after he got in from that difficult shift.

> I knew instantly that it had not been a good night… you could just tell he was vacant. He didn't particularly want to talk to anybody and he didn't want to listen to what anybody was saying. He just didn't want to be there.
>
> He went up to the bath. I left him for a bit and then I went in and he told me about it. It took him a little while to get over that. But he does talk, which does help… He's very good at seeing the bigger picture and understanding

> why he's doing what he's doing. When something like that happens, it's hard knowing that you couldn't do what you're there to do. He knows it wasn't his fault… but when you have to sit and watch events like that unfold it's hard to take really, isn't it? At the push of a button he could have done something, if only circumstance had allowed.

If only. Two words to drive someone mad in the right – or wrong – context. *If only* the clearance had come through clean the first time. *If only* they had another 1,000ft to work with? The calculus of death and regret is an irresolvable sum.

The final words go to Gav:

> It was very weird to go back into a house celebrating Christmas having watched Iraqi soldiers die. And the hardest thing about that is that we were overhead watching with the ability to do something about it. That's the worst – not taking a shot and watching friendly forces die as a result of not taking that shot. It's far harder than anything you might see in strikes or the usual public perception whenever post-traumatic stress disorder is bandied about regarding the graphic things we see.
>
> We got a mandatory session with a counsellor off-station – nothing to do with the squadron – just to sit us in a room to talk through everything, what happened. We all sat down and talked through everything, how it happened. It would be very easy to blame myself because I've got the trigger. However, if the exact same scenario was to happen again I would still abort.

As I get in my car to leave their home I wonder what three years of this lifestyle – five years, ten years – will do to the people involved. I make a mental note to stop whinging to my wife the next time one of my lectures does not go as planned. I also admit to myself that this new resolution will last about as long as my drive home.

CHAPTER 10

INSIGHTS

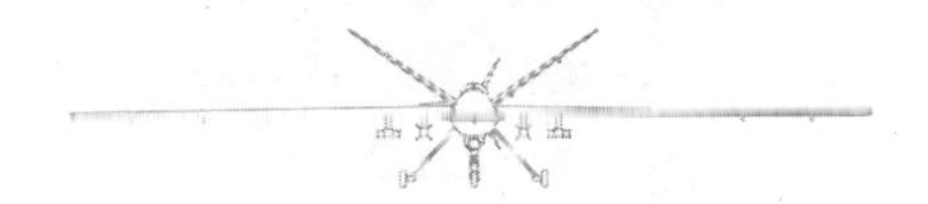

'MY HEART HAS NEVER GONE SO FAST.'

JEFF, REAPER PILOT

Every member of a Reaper crew – whether a pilot, SO or MIC – has one memory or experience that sticks in their mind more than others. Some have several. More often than not, these are mental images of some horror they have witnessed or a recollection of a particular strike. Some of those memories are also humorous, weird or disturbing, and they reflect a whole range of emotions.

I have compiled a range of those experiences below, where crew members speak in their own words. Some are funny, some sad, and some speak of the almost unbearable pressure when a missile is speeding towards its target.

The biggest challenge for most Reaper crew members is taking that step from flying the Reaper in surveillance mode, to taking their first shot – especially the first lethal shot. 'Killing is not easy,' is a common refrain. 'Neither should it be,' is my standard response. Some of the most detailed and intimate recollections of what it is like to make that transition comes from those who have recently advanced from Limited Combat Ready (flying but without taking shots) to Combat Ready (flying and engaging targets with weapons).

More recently, Combat Ready status has been split into two categories so that the jump from benign flying in surveillance mode to lethal shot-taking is less severe. Combat Ready (Restricted) is the first step, and allows shots against static, non-human targets. Fully Combat Ready grants authority to shoot moving HVTs, potentially the most difficult form of missile strike. That's where the real pressure comes in.

Some pilots and SOs have now conducted so many lethal strikes that they are no longer afflicted by the almost debilitating level of adrenaline that typically accompanied their first shot. A small number testify to an almost preternatural calm when about to shoot that has developed over several years and dozens of strikes. The most telling, however, remain the first shots.

The remainder of the chapter leads off with three different 'first shot' experiences – one of which captures an SO's feelings right before, and right after, the first time he takes a life. Those are followed by insights into some of the more bizarre, amusing or random thoughts and events that Reaper personnel experience in what counts as 'normal' in their world. Along the way, Nige provides a lesson on why you should go on a diet before going on jihad, before some final thoughts on seeing the humanity in distant enemies.

COMBAT READY – JEFF

I became combat ready as a Reaper remotely controlled aircraft system pilot – RPAS(P) – about a month ago, and looked at this entry in my Flying Log Book a couple of days ago.

I'd had a supervisory follow-up check to my Combat Ready status just the day before. It's a check to make sure that I was still doing everything properly. We'd done a bit of work-up that day on an HVT, just establishing his pattern of life. Then there was the possibility that things could escalate swiftly towards a kinetic strike. It obviously depends on the individuals and the circumstances.

On the previous day, we had been close to working up a strike. I would probably have been allowed to stay in the seat for the shot since I had just qualified, but I can't say for sure since we never actually got to that stage. But we were getting there: we had all the correct approvals. I picked up a few good learning points that day as well, which I was able to put into place the following day. It had been my first proper time dealing with 9-lines on my own, liaising with those specific JTACs as we discussed various options. For example, we could have brought in other air assets to help us out, to work as a pair. It was a good chance for me to develop – in real time – an understanding how to facilitate the strike between two aircraft.

Having done that work the previous day, on the day in question I was alone as captain in quite a similar situation, apart from being on my own without a 'grown up' sat behind me. Another RPAS(P) came in as the Safety Observer. Everything was spinning up, with all the right processes happening, to strike this particular individual. And then the plan changed. That meant we no longer had the correct approvals for the new plan that we were being asked to carry out. Updates to the 9-line were required and a new update to the approval from the RCH was needed as well. We did not have those in place to enable us to take action right away, which was frustrating. Without those approvals we had to come 'off dry' in the end, without firing.

We were content with the pattern of life that we had established over a couple of weeks, so we knew fairly accurately what time the target was going to be moving. We were confident he would go into a building for a couple of hours, and come out around the same time each day. He often had a child in the car, travelling around with him. We were flying a very tight orbit to make sure we never lost sight of the car, and always knew who was or wasn't in it. This time the child wasn't.

The man drove off. Once we were established in an appropriate kill zone then that was it: a moving HVT was my first shot. It wasn't an armed man in the middle of a field with more fields all around. It was reasonably high pressured – in my mind, anyway. Every time will feel something like that, but there is more pressure on you the first time.

My heart has never gone so fast. I've done a lot of silly things in my time that have raised my heart rate quite considerably, but not like this. The good thing is that I knew to expect it. Because there had been the morning attempt that ended with us coming 'off dry', I was ready. My heart had been beating equally as fast earlier on in the day, so I almost knew what to expect. It felt like it was pumping out of my chest.

I still felt calm in what I was doing, which was good. That was one of the points that came up the day before. When we were looking to spin up an attack, that's when I actually felt like I had a better grip of the situation, and was calm – which is good. That's when you want to be at your most focused, even with the rapid heart rate.

The amount of training that goes into preparing for that first shot certainly worked for me. I had done so much 'work-up', so much strike practice, that when it actually came down to the mechanics of it I was able to work through everything logically and accurately. And carry out the shot without thinking twice about it, I suppose. I fired the missile, but it was a secondary asset, another aircraft, that was doing a 'buddy lase' and using its laser to guide the weapon to its target.

It was up to the other crew to get the missile onto the target; it wasn't the SO next to me doing it. It was strange because I was asked to 'rifle' without seeing the vehicle at the centre of my own screen. When the clearance to fire came, I wasn't immediately prepared to do so. I couldn't correlate the position of the target vehicle with what I was seeing. They had zoomed their camera in so much that all I could see was the vehicle going round *a* corner, and I didn't know if it was *the* corner and *the* car I had planned for. I waited for a wider screen image to confirm the target for myself before firing.

We had a very limited kill zone in which to work. We were counting down the seconds, approaching the point at which it wasn't going to be a viable strike because of collateral damage concerns. However, I had positioned myself so that the weapon was not going to be in the air for too long.

I did my 'Three… two… one... Rifle,' and got the rest of my checks out of the way. Then I was just sitting there watching the screen, knowing that I didn't have very much space left in my kill zone. We'd cleared a very sizable region round the target, so I was confident there was nothing in there to be affected. But there was an east-west running road that *could* have proved problematic if too much time lapsed. We'd checked it and it wasn't a main highway with busy traffic. That was my only concern, to make sure there were no people walking around.

It sounds very strange to say it, but I was sitting there mentally going *Blow up, blow up. Missile, just do your job* because it was getting closer and closer to the end of the available road. I was just willing the car to blow up. It was a time-pressured shot with a reducing number of metres left towards the end. My biggest fear in all of this was not killing the person I had been told to target. I made my peace with that a long time ago. It's the civilian side of things I worry about. That is the worst fear of anyone in this Force, that they injure or kill somebody that they're not supposed to.

So I was sitting, willing the car to blow up. The man inside was a bad man, given what he has been doing. I had no problems with what I was trying to do. I was mainly concerned with the thought. *What if we miss? What will he go on to do – what bad things will happen elsewhere?* There's that added pressure as well. Of course we would 'shift cold' [re-direct the missile into a pre-planned safe area] if necessary, but we did not want to. We worked so hard at this over weeks and weeks to give us the best chance of success in that kill zone without affecting anyone else. So, all in all, it wasn't necessarily the most ideal first shot, but I am grateful that I was allowed to do it.

When you get the intelligence brief on the individual that you are targeting, you know that this is not a nice person. If you did have any problems with pulling the trigger, the knowledge of what these people can do shapes your thinking.

FIRST SHOT – BEZ

My first shot was against a guy firing a heavy machine gun that had been mounted onto a tricycle so he could quickly wheel it around into new positions to shoot. The distinctive horizontal muzzle flash told me that it was a DShk – or 'Dushka' – which fires a 0.50in-calibre round that can go through a concrete breeze block wall a mile away and still kill a guy on the other side of it.

The actual strike happened very quickly. The guy started firing his heavy machine gun at Iraqi forces while we were overhead, watching out for them. We got our clearances and tipped in, got cleared 'hot' and shot. I'd love to know what my heart rate was when the trigger was pulled; I reckon it was pushing one hundred and seventy to one hundred and eighty beats per minute.

The machine gunner heard the weapon in the last couple of seconds of its flight, turned and tried to run. But it was in vain. The Hellfire impacted two or three feet behind, killing him instantly. There were other Daesh guys firing smaller weapons nearby as part of the same attack. As soon as the blast went off they instantly turned and ran away, ending the contact. As a crew, we went right into our BDA and were paying close attention to the detail of what happened.

A few minutes later, as we continued our surveillance of the impact area, we watched a little white vehicle pitch up – it looked like an old Mazda Bongo truck. Three guys jumped out. The first two grabbed the gunner's body, loaded it into the truck and climbed back in. The third guy scooped up a sizable part that had been dismembered by the strike – it was a result of the way the fragmentation pattern of the Hellfire had hit him – and plopped it onto his chest. They then whisked the body away, no doubt for a 'martyr's funeral'.

It was a pretty grisly first shot; we don't always see the aftermath in such detail. After the event, and once I was out of the GCS, a few of the guys pulled me aside and asked if I was OK.

It was only afterwards, on a treadmill a few hours later, that I realised that the only thing I was beating myself up about was what I could have done better, could have been a better operator. The idea that I, as part of a crew, had taken a life didn't trouble me. Somewhere, there was an Iraqi soldier who might just get to go home to his family that night, rather than die from a 0.50in-calibre bullet from that machine gun. Plus, that city was ever so slightly closer to being liberated from Daesh. I have literally never lost a moment's sleep over that strike, and doubt I ever will.

FIRST SHOT, BEFORE – PHIL

As a Limited Combat Ready SO, it is frustrating at the moment not to be able to be involved in taking the shots where I have done most of the build-up. We train on operational sorties day-in and day-out. We practice numerous simulated weapons events. And those weapons events are testing: we test almost every conceivable option.

In those practices, the instructors will challenge our knowledge by deliberately leaving out a key bit of information that we require. Or by giving false information in the simulated scenario that we need to recognise and not be caught out by. They do it so that we will pick up on that false information or omission, thereby proving our knowledge and understanding of our rules and regulations. And in knowing those rules and regulations – that are there to protect me – you go through all the permutations before you get involved in a strike event, simulated or real.

In a real event, when I have worked up all of the information and necessary permissions, to have to get out of the seat is very frustrating. I think, *I could do this!* I wouldn't say it's the easy bit, it's not by far, but I think I could handle it now.

When I get out of the seat to let the fully qualified SO take the shot, I will stay in the box to learn and understand what I can. I do that so that when I become Combat Ready and I am in the seat, I understand everything that happens. There's an element of training and learning that takes place from every single weapon event, whether you are watching it or doing it.

I'll stand behind the instructor, or that Combat Ready person who is now in the seat. Whilst there is a huge weight of responsibility off my shoulders, I am still in the 'big brain' or 'capacity' seat behind the pilot and SO. The extra person in that observer position does not have the pressure that goes with taking the shot and can often more easily see what's happening across the big picture than the crew who are deeply involved in the detail. There will be a reason why the instructor has taken a particular approach and I want to learn it all.

Once the event's all over, every strike is monitored and recorded, then analysed and debriefed by a weapons officer. It's all about the learning.

At the end of our first discussion, Phil indicated that he would be willing to talk to me again. So I invited him to contact me after he had been involved in his first weapons event. Around three weeks later he emailed me the following: 'Since our last meeting I have now had my first engagement, what a gut-wrenching feeling. As and when it's convenient, you mentioned you'd like to interview me again, which I'd be happy to do. Phil.'

FIRST SHOT, AFTER – PHIL

My first shot was an interesting experience. Having done all the practices, and it doesn't matter how much you practise in benign scenarios without an actual outcome, when it happens for real, it's a very different feeling.

The task that we had was complicated. I wasn't supposed to have a complex shot for my first engagement, but I was in the seat as the situation developed, and I had qualified as Combat Ready. I just happened to be on that operation at that time. We'd been given a clearance from the JTAC to engage a particular target – a static target with enemy forces moving about in the vicinity. The reason it was complicated was that there was also transient traffic and people. It wasn't as easy as, 'One, two, three… go': a nice, clean 'vanilla' shot against a static target out in the open.

After we had been given our clearance to fire, we sat for nearly an hour, prepared and ready. Several times we started our run-in to fire, and then called 'off dry' and aborted the engagement. It needed a lot of tactical patience, mainly because of the transient people and traffic that were entering and leaving the target area – huge potential CIVCAS.

The whole situation quietened down a little bit, giving us an opportunity to check and re-check what was happening on the ground. In the camera's various fields of view, from wide to narrow, I scanned for weapon clearance distances and transients who might be at risk. At the widest field of view it was all clear and we had our opportunity to engage. We were using a Hellfire missile, and the aim was to destroy a checkpoint and the enemy fighters manning it.

It was a successful engagement but even after all of those benign practices, now it was for real, my heart started pounding from the point of being 'cleared hot'. For that hour I had been on tenterhooks, nervous. Especially when we ran through our checks and the captain said, 'laser'. This part would be down to me.

The laser was switched on. I was doing the lasing – using the crosshairs on my screen to aim the laser at where the missile would hit the target. My left hand felt like it was almost in a death grip. I was so tense because there are things that can go wrong if I press buttons in the wrong order. For example, the missile might not arm. So I made doubly sure everything was correct, with the laser on. It was almost like a white-knuckle ride: you're holding on so tightly because you don't want to make a mistake.

Once the missile was off its rail and heading towards the target, the pressure was on me for the duration of its time of flight. You're under pressure not to move the crosshairs and cause a 'miss'. You have to keep the crosshairs on the target while the aircraft is moving and your target is moving in relation to the aircraft while, at the same time, looking out for potential transient people or vehicles. If any transients had appeared I would have shifted the laser pointer – and therefore the missile – into a pre-determined clear area, avoiding any CIVCAS. But there were no such interruptions.

We struck the checkpoint; the missile went exactly where it was supposed to. It wasn't a 'normal' feeling, it was an abnormal thing to do. Even talking about it now I feel my stomach turning. It was gut-wrenching that I've now taken people's lives. It's a weird feeling.

Throughout the whole hour, the adrenaline was fluctuating a bit, but it was imperative that we hit that target. On those couple of opportunities to engage where we came 'off dry', without firing, I guess if I was wearing a heart-rate monitor it would have been pounding nineteen-to-the-dozen. I was in that state for the whole hour.

At the time I was thinking about how they were enemy forces and the bad things and atrocities that they had been conducting, that in taking those lives we'd be saving many others and prevent them from suffering. Even then, it just feels so abnormal to go and take someone's life, even when it's an individual that deserves it in a military sense.

When you find you've taken a couple of lives with – dare I say it – a 'clean shot', that's fine. But if a couple of enemy fighters are wounded and you have to go back and assess the damage to create an accurate report, that's almost harder to deal with. There's the awareness: 'We did that to them, and they're suffering now'. Weird feeling.

QUITE A FEW MONTHS LATER – PHIL

I drove home last night after a very busy day, trying to unwind. Trying to clear my mind of the day's events. I found myself deep in thought, distant and cold. For the first time, I was thinking: *What have I become? No one really knows 'me' anymore*. I find this slightly disturbing.

I know that what I'm doing is protecting and saving

lives, but at the same time it's so hard. It's lonely. I can't tell my partner what my day at work was like. This morning I'm out walking my dog, through the soaked wet fields, no one for miles around. I find myself sat down talking to my dog for comfort. Her unconditional love keeps me strong.

It's a weird life.

And now for some of the more bizarre experiences and observations...

IRONY. AFGHANISTAN 2012 – ROSS

During the crew brief we were informed that our task for the day would be to take over from another aircraft, which had been watching an IED 'emplacing' team sleep outside a compound all night. The IED team had been observed laying an IED in the road the previous day but they could not be engaged due to concerns about collateral damage while they were mobile on a motorbike. As was commonplace, they had bedded down outside an unassociated civilian's 'guesthouse'.

Once on task we 'match eyes' – make sure we are both seeing the same picture – with the departing aircraft. We gain a positive handover of the IED team, still sleeping adjacent to their motorbike. Then a 9-line was issued. Knowing what it entails is a unique scenario, one that is hard to explain. Personally it triggers a higher level of focus and starts a run of adrenaline that lasts throughout the event and can sometimes, exhaustingly, last for hours.

The IED team consisted of two adult males – a specialist who constructs, transports and lays the device, and his assistant who drives the motorbike, digs the hole and watches out while the IED is emplaced. As expected, once the sun was rising, they got up and made breakfast. We watched and waited – this was a fairly commonplace event. As the MIC, I had already familiarised myself with the roads and tracks in the area and had all the intelligence about where the IED team had been since their identification. The pilot and SO were briefed and up to speed. My thoughts now were purely on when and where we could safely engage the two men when they set off. We waited and discussed the options.

Once their breakfast and ablutions were over, the IED team gathered their things and set off on their motorbike. Another hike in focus and another surge of adrenaline. Striking a motorbike as it transits through the Afghan countryside was extremely challenging. The photographic mapping used to compare where the enemy are and where they are going is extremely useful. It gives you an idea of where villages, compounds and open stretches of countryside are, but they do not take into account other vehicles, people, pop-up bazaars, herds of sheep or newly constructed compounds.

A suitable strike zone was identified, with a suitable 'shift cold' safe area, should it be required. It was necessary for a 'zoom out and scan ahead' to be conducted to ensure that the planned strike area was safe. So, once the direction of travel was determined and the strike area was deemed free of non-combatants, the mandatory scan ahead occurred, looking for non-associated persons or vehicles. The scan kept the motorbike in view at the rear of the screen so as not to lose the PID. When we were content that the strike zone was clear, a 'trigger point' was selected, which all of the crew knew was the point where the AGM-114 Hellfire missile would be fired. The average speed of the motorbike was known, so with a known missile time of flight being vectored, this gave an accurately predicted weapon impact area. This area was seen to be clear.

As usual during a weapons event, the chat among the crew was limited and deliberate: 'Final zoom out… Trigger point reached – now… Everybody content? Three... two… one… Rifle.' The missile was away.

Predicting the unpredictable is obviously difficult. And five seconds into the missile's time of flight, the motorbike stopped at a previously unseen grass hut. The grass covering completely camouflaged the hut against the background field, making it impossible to see during the camera 'zoom out and scan ahead'.

A decision had to be made. Will the bike set off again and be clear of the hut or will a 'shift-cold' have to happen? The missile was in the air and the impact time was counting down – twenty seconds… fifteen seconds to impact – and still no sign of the motorbike moving. There can be no risk to a potentially unseen person in the hut, so two of the crew called 'shift down' concurrently. The SO moved the cross hairs to the pre-identified safe location in the middle of a field at the bottom of the screen. The impact threw up a huge cloud of dirt and dust.

Immediately after the weapon 'splash', the pilot directed the SO to get eyes back on the motorbike. There was a sense of disappointment, but the incident was not over yet. The passenger, on hearing the missile impact, jumped off the motorbike and ran towards a compound beyond the hut. The rider set off on the motorbike and headed along the road at speed.

The pilot directed the SO to follow the motorbike with the camera.

For some reason, after one hundred metres or so, the motorbike came to an abrupt halt. The rider immediately looked over his shoulder towards the area of the compound, dropped the motorbike and started running in that direction. I directed the SO to look in that area since something had obviously happened to cause the bike rider to stop.

As the camera view moved towards the compound, a plume of smoke and dust filled the screen. When the plume passed, we could see an adult male lying on the ground, just beyond a gap in a hedge, and adjacent to the compound.

The prone man was recognisable from his clothing as the IED passenger from the motorbike. The rider arrived and rolled him over. It was obvious that he was dead as the result of a catastrophic explosion. Our first thought was that he had been struck by another missile, but there were no other aircraft in the area. So we were initially confused about what had just occurred.

As we examined the area he was in, it became apparent that he had run around the compound, gone through a gap in the hedge and tried to return towards the road to catch up with the motorbike.

Gaps in hedges were a favourite Taliban location for laying IEDs, as they are natural 'choke points' that Allied foot patrols have to pass through. Fortuitously for us, and ironically for him – the IED specialist who evaded our strike – he had triggered an IED intended for coalition forces and was now lying dead, possibly by his own handiwork.

THE TECHNICAL AND THE DOG – BEZ

We were coming off task after conducting overwatch of an Iraqi city, when we were tasked to investigate and subsequently interdict, i.e. strike, two 'technicals': basically, pick-up trucks with a weapon mounted on the flatbed. They're relatively cheap and easy to build, but when operated well they can cause all sorts of hell for guys on the ground and for low-flying aircraft and helicopters.

We got 'talked-on to' the vehicles' location by another ISR aircraft, which didn't have weapons on board. Sure enough, we found them parked in two separate car shelters. We did our standard look around for collateral damage concerns. That's when I spotted a small glow moving around the building while scanning with the camera in infrared view. It took me a few moments to figure out what it was. I could tell it wasn't a person because of the way it was moving. And it wasn't something blowing in the wind. It was a dog, just sniffing around and minding its own business, now right next to the target vehicle.

We did a practice run, got our aiming mark and weapon fusing correct. At this point, two armed guys came out of the adjacent building and got into the truck. Realising that they were about to leave to conduct some sort of Daesh action, the pilot spun the Reaper back out, got some distance and then tipped in – putting the aircraft at the right distance and correct angle – to strike.

I zoomed the camera out to check again for any collateral damage concerns (transient vehicles and people), and then zoomed back in. I couldn't see the dog. I prayed he had scampered off and wasn't anywhere near the technical.

We got 'cleared hot' to shoot and put the Hellfire straight into the weapon on the truck.

Seconds after the weapon hit, I saw a hot spot come screaming out of the back door of the building. It was the dog, which, despite having had the shit scared out of him, was completely unharmed. It ran to a nearby house and sniffed around outside for a few moments before continuing on its merry way.

I ended my stint in the box ten to fifteen minutes later, and returned to the Operations Room. A few of us watched the video over and agreed that the technical had been destroyed. I also confirmed to myself that the dog was OK, much to my relief. It was only then that I realised that I didn't give a moment's thought to the two guys in the truck. They were dead. But the dog had made it, and to me that was the important bit.

It put the whole thing into perspective for me. My own interpretation of the rights and wrongs of taking life in conflict is predicated around innocence – or lack thereof. That dog was just that – a dog. Innocent, minding its own business. It'd probably never hurt anyone in its life. The two guys in the truck were out to hurt people on behalf of a tyrannical organisation whose capability and reputation revolved around inflicting ultra-violence, oppression and fear on completely innocent men, women and children.

BULLIES – RALPH

We were watching a building for some time, hoping to learn more about the guys that worked out of it. It was suspected of being a main IS hub in the area. It was what I call a 'low stimulus trip' – not much going on, not much sensor pod manipulation, just stare at the building and back the pilot up in some good aviation. Boring stuff.

As I sat sipping my coffee – the perk of being a Reaper operator – I noticed on-screen that in the street outside a group of kids were playing. They were probably between thirteen and eighteen years old, and pushing each other around, standard kids' stuff. Then I realised that, actually, they were pushing just one kid around. Then I realised that this kid wasn't moving around normally, and that his coordination looked all to pot. Finally, it hit me: this kid was disabled. It looked like cerebral palsy. And he was getting bullied by the other kids.

I got quite angry watching this, and willed the poor lad to turn around and smack one of them back, to defend himself. But he just didn't have the coordination to do it. And he didn't have to.

A few minutes later, a fully grown man who had been leaning out of an upper balcony of the target building looking down on the street came charging out of the door. He walked over to the biggest bully in this group, and gave him a slap the likes of which I had never seen. I mean, I was astonished that this lad was standing after the backhander he got. Instantly, the bullying stopped. The big bully held his face, his pride as bruised as his cheek would no doubt end up.

The kids then walked over and sat on the wall with the young disabled boy. One of them ran off and got some food, which was shared amongst them all. When we came back to the same building a few days later, we saw the same group of kids, but not one dared bully the little disabled boy – he wasn't pushed or shoved or teased. They all played together. Proof that sometimes, despite what mother told you, violence is a valid answer.

HOKEY COKEY – GEORGE

One moment which brought humour to everyone's day occurred when we were flying a mission over northern Iraq. We had located an individual who appeared to be a Daesh field commander, who spent his afternoon driving around the area checking-in on his troops. We duly followed his movements and kindly noted the enemy dispositions. He was accompanied by two armed subordinates.

They arrived at a small, single-storey building, with no windows and just one door. The commander seemed keen to gain entry. Unfortunately for him, the door was locked. He evidently ordered his men to break down the door.

We watched as the first one tried to kick the door down, but each attempt failed. He was then joined by the second man, whereupon they placed an arm around each other. Side-by-side, they took several steps back. They then took several steps forward, before trying to kick the door at the same time. This again was unsuccessful. So they repeated the backward steps back and attempted the joint run-up once more.

Despite their attempts to coordinate, it looked like some Daesh version of the Can Can or the Hokey Cokey. After a few more minutes of failure, which had brought chuckles from our crew, the commander seemed to issue a stern bollocking and ordered them elsewhere with pointed finger.

I second-guessed that they were now likely on the prowl for something substantial to break the door with. Sure enough, they returned a few minutes later with a ladder. They then proceeded to attempt to use the 12ft long wooden ladder as a battering ram, a tactic which was doomed to fail. Needless to say, this was even less effective than the previous synchronised kicking. The result was a broken ladder and the two individuals falling over, like an episode of the Chuckle Brothers. At this point we received a message from the CAOC, who thanked us for providing them all with much amusement, as they had screened the entire episode on the giant video wall for all to see.

Perhaps the most revealing aspect of Reaper operations is the way in which they redefine 'normality' for those involved…

NORMAL – NIGE

By the fourteenth shot it was becoming very ordinary. I thought about this a lot after my tour ended. At the time I didn't. It became very, very 'normal', especially when we went through a time when we did a lot of shooting. But I think a lot about it now, honestly. I know a number of the other guys think the same. I think the term is 'normalization of deviance'. Not 'deviance' in necessarily a bad way, but you're deviating from the norms of society. Our frame of reference on the Reaper Force, like that in any fighting force, moved away from every-day normality. When you're on the Reaper Force, what is 'normal' is not normal. But you think it is.

It's not just the kinetic strikes. I remember an example from 39 Squadron. Three separate events that we had been involved in over a number of nights were widely reported in the world's press. For us, this was absolute normality. The lives and work of most people in the 'normal' world are never reported in the press. And on that occasion, 39 Squadron had been involved in three separate events that made news around the world – in three nights.

CHILD SOLDIERS – ETHAN

Not long ago, my friend and I were chatting about what we had seen in work a few hours previously. We had observed a child soldier in the company of a few other, older, armed males. He was quite clearly a child. It was also clear to see that he was under the control of the accompanying adult men.

My friend and I were standing outside my accommodation having an exceptionally serious conversation about some of the 'what ifs' associated with this child. We spoke about the potential impact he could have on the friendly forces as a combatant with a weapon. We also spoke about the potential impact the UK Reaper Force could have on him. Then there were the rights and wrongs of firing or not firing on a child. We had very furrowed brows and were rather glum, to say the least.

Just as we were finishing the conversation, a group of RAF administration support staff arrived, full of the joys of spring. They were waxing lyrical about what a good day they had had down at the local water park. My friend and I knowingly looked at each other and smiled, acknowledging how different our two worlds really were. It still makes me smile just thinking about how the admin staff didn't have a clue what we were having to go through, and the scenarios we were wrestling with.

TRANSITIONS – RORY

I enjoy the core job of what we do. Being able to have a positive effect on the ground in an operational theatre every day, and going home to my family at the end of my shift. It's an odd situation, but it's the perfect situation for me. The car drive takes about thirty minutes. I don't even think about that adjustment from operations to home, because it has been going on for so long now. I like the drive because it does give me that break.

Recently, I have been finding it more and more difficult to switch off when I get home. Definitely. I don't know if that's because I'm ready for a break after five years. I don't think you ever fully make the break between work and home, and home and work – it is blended so much. Yesterday was a prime example. I had been doing office work upstairs on the squadron when I was called to go straight downstairs to the box and take a shot for someone because they were not qualified. When that was finished I came straight upstairs and there, on my phone, my wife had sent me a video of our son walking around and playing with his new toys.

Mentally, I had moved instantly from an operational video feed and a shooting scenario to a family video. She also texted me and asked, 'What do you fancy for dinner tonight?' I thought, *I don't know, I don't know. I can't think about that.* It seems such a trivial thing, but I couldn't think about the food.

FOLLOWING A PROCESS – NIGE

This memory is massively random and I run it around in my mind a lot. We were tracking a load of guys on the ground who had engaged Iraqi forces and then retreated. They were moving tactically, spacing themselves, deliberately making it difficult for us to attack them with a low-collateral weapon like the Hellfire. I watched them for some time and I actually worked out who was the guy who was most overweight. He was somewhat distinctive and it was quite simple.

I decided that if and when they started to run, when they dispersed as was usual when they heard the incoming rocket, he was going to be the guy I would go for. I'd watched them for some time and could tell that he was less fit than the rest of the guys. So I selected him.

I don't think it troubles me – it's quite sensible in a cold, objective way I suppose. But I actually went through a process where I selected one guy out of five or six. My thought process was: *He's the one I'm going to go for because he's the easiest. He's likely to be the slowest*. It was just absolutely cold reasoning at the time.

In the car on the way home I reflected on the engagement and it struck me: *Crikey, what does that say about me? I made a decision to maximise the effectiveness of the weapon and kill a guy based on his weight*. In the event, the men grouped together and we got three of them with one strike. My thought process throughout the engagement was very clinical. That was especially the case in my first six or seven shots. When I got into the situation where I was shooting it was, absolutely, 'I am following a process here.'

I exercise most days on a running machine. In the beginning of my tour, I'd be running and going through all of the motions of taking a shot. I did it so I knew that if the pressure got high, I could rely on going back to almost rote-learned stuff. I mentally went through the zooming in and zooming out with the camera, the talking, everything. On that running machine in the gym near where I lived – with nobody else in the place – I'd be going through all the checks. And doing it again and again and again. I must have done around a hundred shots – in my mind and with my hands – on the running machine, prior to my first one. Just going through the process.

If there is one suggestion that is likely to rile a member of the Reaper Force, it is the idea that they are emotionally distanced from their enemies. I suspect the job would be much easier if they were. Everyone has a unique capacity for empathy – some more, and some less, than others. If you are going to recognise the humanity in someone you kill, or plan to kill, there will be an emotional cost at some point…

SEEING THE HUMANITY IN AN ENEMY – TOBY

One thing's really stuck in my memory. It has affected me. Even my mum said, when she came to see me in Vegas, that I was changed.

One day we were just randomly scanning an area of IS-held territory that spanned Iraq and Syria, just being steered by the Supported Unit. We saw this high-loaded truck and there was an armoured personnel carrier (APC) on the back of it. A huge transporter with a big, huge truck in the front. And it had an outrider car going very fast. I would say it was the lead car. The other vehicle was just following. They had stolen that piece of kit and wanted to get to another area in a hurry to fire it against Iraqi troops.

So, what we did, we struck the first car. Clean hit – happy with that. At that time, viable targets were defined simply as ISIL or ISIL-associated vehicles – anything was viable that fitted those target descriptions. We were happy that they were ISIL fighters. We'd done all our checks, confirmed through intelligence sources it wasn't friendly kit and there were no friendlies in the area.

Having struck the first car, the big truck behind it slammed its brakes on and stopped. Two guys ran out, which was fair enough. They ran out of the cab and left the truck, with the huge APC on the back. It was badly covered by tarpaulin – you could see that it was an APC.

So they ran into the mountains. When we were content that they'd run off, we put a Hellfire into the APC and blew it up. Everybody was happy. We did some further scanning around then we struck the cab of the truck. Another good shot. But these two guys came back to inspect what happened. I felt sorry for them. I felt some association with them because my dad used to take me for drives in trucks where he used to work. I felt an affinity with them.

Then we were 'cleared hot' on them. We rifled them [used a Hellfire]. They heard the sound of the missile as it approached, then one of them threw himself on top of the other. And that's just stuck in my head. I don't know why it's affected me. I don't know if it was the affinity with my father and the truck. I believe they were father and son because of the way the one threw himself on top of the other. And they were both lying there, dead. It's just something that's always stuck with me. They were just men like me.

And then I'd come home after the shift and I'd be handed a baby. Thirty-eight minutes was my drive home. I've done conventional tours overseas where you'd get a cool-down period on the way back. You'd stop at Akrotiri in Cyprus. To get home from Afghanistan or Iraq was a two- or three-day process. So there was that time to think, to start to adapt.

You'd even get shown the video about how you might react when you got home after months away. For example, you might get a bit 'fighty'. But with Reaper you're right back home every day.

SCARED – ZARA

It was right that we took the shot. We'd been chasing a group of IS fighters with weapons who were shooting at friendlies – all day. It had been a busy week. We'd been watching this team, and we tried to get them as a group but we couldn't. They split up and we were going after them one at a time.

One of the combatants realised that we'd hit his mates and we were following him. And we watched him being sick, because he was scared. We were then 'cleared hot' to use a missile against him, and that strike will stay with me forever. We were on opposite sides, and what they were doing was horrendous. But still, they must get scared.

'SHIFT COLD' – ERIC

I just go into overdrive with enthusiasm about the job we did, the expertise, that coolness. We were *so* careful. When I saw protestors' signs with 'Baby Killer' on them it used to make my blood boil. We were the most careful, precise, thoroughly professional unit, and it was never advertised. The amount of 'shift cold' that happened shows that.

My job as a MIC was to make sure that we had a 'shift cold' location, somewhere the SO could divert a missile to while it was in the air if any risk to civilians emerged during its flight. I would say, 'OK, if we're going to go "shift cold", we're going to go "screen south".' That would mean a piece of land at the bottom of the screen.

We'd work out what we would do before a shot was fired. I used a clock method. For example, 'We'll go six o'clock into that field. It's walled and there's nobody in there.' In other words, the SO would redirect the missile into a walled field at the bottom of the screen in the six o'clock position. And we would do that time and time again.

By the time I left by 2012, we'd had four hundred missiles fired or bombs dropped. Over that whole time we had one bad collateral damage incident where civilians were killed. After that particular incident it was sombre. I came in to the squadron the next morning to take over from the crew. I was speaking to a couple of the MICs involved and they were *absolutely* devastated by the thought that they were involved in an incident that cost innocent people their lives. *Devastated.*

Doing what we did affected people. As soon as I left those gates at the end of a shift, it was all behind me. I would be travelling through the desert, over the hill, and then see the lights of Vegas – a different world. I tried not to let it affect me too much.

For the sake of balance between the two RAF Reaper squadrons, I should point out that the insights in these interviews come from people with a fairly even split of experience between 39 Squadron near Las Vegas and XIII Squadron in Lincolnshire. Each location holds a number of attractions for the people who live and work there.

Las Vegas just seems a much more fun place to be than Lincolnshire. In the unlikely event that anyone on 39 Squadron actually has any time or energy to take themselves to the Las

Vegas strip, a mind-boggling array of delights await. From casinos, restaurants and strip clubs to concerts, stage magicians and, well, more casinos, the energy is undeniable. Things are big, bright and loud. Even the billboards next to the eight-lane roads that circle Las Vegas are amazing. Lawyers, plastic surgeons and dentists – all with perfect, straight white teeth – smile down from giant posters and offer to make your life better in some way.

As someone who loves Lincolnshire, having lived there for over ten years, I should point out that it is not without its merits for those on XIII Squadron. On the entertainment side, the annual Heckington Village Show features events like the 'Most Attractive Donkey', 'Largest Rabbit' and 'Longest Carrot' competitions. The main roads around Lincolnshire are, admittedly, less impressive, and signs that warn drivers not to kill badgers. Judging by the amount of roadkill around the county, it might be more effective to warn badgers about the drivers.

While it is easy to point to obvious differences between the geographical locations of the two squadrons, there are practical implications for those who carry out Reaper operations. Crews on both sides of the Atlantic regularly witness distressing or gruesome sights. Even more so in the intense fighting on the ground in places like Mosul, Raqqa, Homs and elsewhere through 2017 and into 2018. If it happens to a crew on 39 Squadron at Creech Air Force Base, and if they finish work at some anti-social hour, they will have the option of all-night coffee shops, restaurants or bars in or around Las Vegas where they can go, talk and wind down before heading home. If you finish at an anti-social hour on XIII Squadron at RAF Waddington, that option is not there. But they do get home much more quickly than those at Creech, living closer to the base.

Now, to soldiers, Marines or others who have fought on the ground in Afghanistan or Iraq, this might seem like quibbling over a luxury beyond their imagining. I would just ask them, or anyone else, to think about the thing that I dwell on: how would they feel about kissing their spouse, partner or children within hours – and sometimes minutes – of taking one or more lives and watching the aftermath in great detail, repeatedly, over months and years? And that aftermath can include watching the response of young children to seeing a horrifically dead parent. After all, it's not like the Army, Marines or Special Forces have all successfully reintegrated back into family life after they have been away on operations.

CHAPTER 11
MEDALS

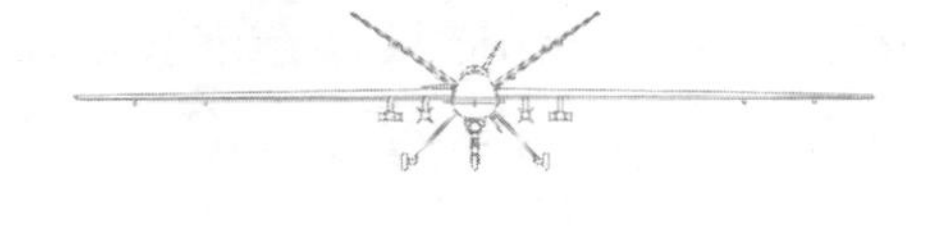

'9-LINE APPROVED FOR THE SNIPER ON THE TOP LEFT OF THE BUILDING.'
RED CARD HOLDER

Tuesday, 20 September 2017. Today. Time will tell if this proves to be a significant day in the history of the Reaper Force or whether it will just be one of the many days when bigger news stories push it to one side. This morning, three things happened. The award of campaign medals for Operation SHADER in Syria and Iraq was announced, with none being awarded to members of the Reaper Force. However, the Defence Secretary, Sir Michael Fallon, did publicly raise the possibility of RAF Reaper personnel being awarded some kind of medal or public recognition for their work on operations in Iraq and Syria against IS in the future. And an RAF video clip of a missile strike appeared in the media.

Nobody was keen to point out that campaign medals are regularly awarded to military personnel who might technically be deployed on operations but who stay firmly protected in hardened buildings and go nowhere near danger. In the First Gulf War, campaign medals were even awarded to personnel based in Cyprus, and it is rumoured that the RAF windsurfing team, which was marooned there at the outset of the conflict, qualified for the medal. Having been based at RAF Akrotiri in Cyprus myself during the Iraq War in 2003, I can testify to the long hours that were worked there in direct support of military operations. Meanwhile, Cyprus remained a popular destination for holidaymakers during both the 1991 and 2003 Gulf Wars, oblivious to any real threat to the island. Statistically, some deployed personnel face less chance of being killed than someone riding a motorbike in Lincolnshire.

Two issues dominate the debate about medals for Reaper crews: their physical location and the fact that they kill from afar; and the historical tradition for awarding medals for geographically specific campaigns and the associated 'risk and rigour' in theatre. The Reaper challenges those assumptions in many ways. One of the ironies of that day was that the video released by the MoD did not feature a Typhoon fast jet strike, but one launched by a Reaper. It showed a public execution being disrupted by the unexpected intrusion of a precisely delivered Hellfire: 40lbs of high explosive delivered at hundreds of miles per hour put a bit of a dampener on a public beheading.[30]

The video clip was used by a number of media outlets to provide a backdrop for the story about Operation SHADER medals. However, few of them highlighted the paradox that the Reaper personnel responsible for the missile strike – and many other strikes that have saved numerous lives – did not receive official

30 The video can be seen at https://www.youtube.com/watch?v=-LvWMZKusEs, accessed 20 December 2017.

medallic recognition. So let me describe that particular missile strike, which was much more complex than it appeared…

For the most part, the pre-flight briefing was much the same as countless others: the usual cautions, caveats, situational awareness update and checks. The intelligence picture in the operating area in Syria was being continuously enhanced. Meanwhile, IS jihadists were enforcing their political and religious will across the lands and people they had seized. They were not universally welcomed by Syrians and Iraqis, who did not want to be part of any supposed caliphate. Human intelligence – information from people on the ground – was steadily trickling its way to the coalition intelligence-gathering agencies to augment the information that could be gathered from the air.

The focus the day of that missile strike was on the town of Abu Kamal and the surrounding area. The intelligence that the squadron had received had been vague, but there were numerous indications that there was 'badness' around. They were to help build the intelligence picture by watching out for any obvious signs of IS controlling the area.

The first crew into the box got the mission under way. They expected a standard, routine reconnaissance trip but there was one slight complication. They were an Op 2 crew, which meant they did not hold the necessary classification to fire weapons, though there was an Op 1 crew – Gav (pilot), Brodie (SO) and Ricky (MIC) – in the building who were qualified and available to take over if needed.

Gav and Brodie were under extra scrutiny. As well as flying on operations, they were doing so as part of their Qualified Weapons Instructor (QWI) training course. They didn't just have to be good at their jobs. They had to be good enough to also teach others how to be good at conducting weapon strikes.

So on top of flying a basic sortie, they also had additional actions to carry out. Throughout, everything they did and said would be assessed by one of the most experienced pilots ever to fly a Reaper, someone with thousands of hours of fast jet flying as well.

After a long transit, as their Reaper approached the day's operating area, the Op 2 crew double-checked their briefing notes and any updated information that had emerged from the CAOC. During the approach to the town of Abu Kamal they crossed what was still, on school maps at least, the Iraq-Syria border.

When IS seized a number of these cross-border areas in 2013, they made a point of demolishing old border posts and the sand barriers that had been put in place between the two countries. Then, on 4 July 2014, Abu Bakr al-Baghdadi announced the creation of what he called the Islamic State and declared himself to be its Caliph, calling on Muslims everywhere to submit to his authority. In that same speech, he rejected the idea of a border between Iraq and Syria, arguing that it had been artificially created in the Anglo-French Sykes-Picot Agreement of 1916. To reinforce his point, IS released videos on YouTube showing its workers and fighters driving bulldozers through those old fences and barriers. These cross-border areas, and the major towns on either side, now provided key hubs for moving people and resources – especially weapons, cash and oil – between them.

The Op 2 SO was intermittently zooming the camera in and out, focusing in on any activity that caught their attention. They spotted a number of what were, in effect, a cross between foot soldiers and police: low-level IS fighters who were not of too much interest while they were standing around in the street as life went on around them.

As the Reaper reached Abu Kamal the crew was on high alert. The intelligence had not been too specific but one or two important IS members might appear in the area. Something might be about to happen. With the current intensity of operations against IS, something was always on the verge of happening.

They could see the housing and population density increasing as sparse countryside met town development. Isolated houses gave way to small settlements, connected by dusty roads to the larger conurbations. Occasional trees provided shade from the sun along the way, with every compound seeming to have one or two of their own.

Initial observations revealed the town to be typical of many that had been seized by IS over the previous year. Some buildings had been damaged and there was the occasional vehicle that had been burnt out or blown up – not always easy to tell which from a rusting shell. There had been sporadic fighting here, but not the full-scale urban warfare and infrastructure destruction that would be seen later in Raqqa and other major cities.

Regardless of IS's political and religious aims, and some of the medieval violence they used to shock, threaten and entice their different audiences, the lightly armed rapid manoeuvre warfare they had used to seize vast tracts of land against much larger forces was impressive in purely military terms. Even better, from their point of view, fleeing Iraqi soldiers had left behind weapons, vehicles and other resources. So, control of the border allowed seized Iraqi Army vehicles – some of them high-end goods provided courtesy of the United States – to move around freely. At least until American-led coalition air power entered the fray.

Mostly things looked normal. Perhaps too normal. From the air, of the many things that the experienced observer knows to look for is patterns where there shouldn't be any, especially movement of people. And in one of the side streets people only seemed to be walking in one direction, in dribs and drabs.

The Op 2 MIC asked for the camera to zoom out a bit to show a few more streets. As they approached a main crossroads, the eyes of the crew were immediately drawn to a slowly gathering crowd. More armed men could be seen. The increase in number and concentration of IS weapons on the ground meant that there was the possibility of their mission escalating from reconnaissance to the potential use of missiles or bombs from the Reaper if things got violent.

Gav, as a duty Op 1 pilot, got the call to get down to the GCS right away and take over. Brodie and Ricky would be right behind him. They were going in blind. They had not been watching events develop and each needed a quick briefing from the person they replaced. They went from planning room to the front line, in seconds.

As they quickly assessed the situation, the trainee pilot stayed in the GCS to learn what she could from watching Gav and how he flew the mission. The rapid development of events meant that the usual airspace control was having to be accelerated to make sure other friendly aircraft did not end up in the wrong place if missiles or bombs were used.

'This looks interesting,' muttered Gav. Ricky the MIC asked Brodie to zoom in on the people walking slowly towards the crossroads. 'Are they being shepherded?' It did not take long – seconds rather than minutes – to spot the black-clothed, rifle-wielding jihadists herding locals to the same place.

'Going round.' Gav started a slow, circling turn.

They watched a crowd gathering around the main intersection. Several buildings surrounded the area, with one taller building overlooking the scene below.

'All that we're missing is a roundabout for a public execution,'

Gav half-joked. Both Reaper squadrons had already seen more executions than they cared to think about, and roundabouts had featured in several of them. They provided a good central viewing point for the crowds and usually had multiple roads leading to and from the vicinity. People could arrive and depart quickly and easily.

The scene unfolding below started to look more coordinated than it had first appeared.

'Confirm that we have at least one armed individual directing people from the right on the screen. They are heading to the crossroads.' Ricky had been able to rewind a few seconds of footage and analyse it in greater detail.

'Checking the next road.' Brodie moved the camera to another of the lanes where people were slowly drifting to the central crossroads.

'Confirm more armed men,' added Ricky.

'Let's stick with the intersection and see what's going on.' Gav had seen this kind of set-up before. 'Look at the figure in the centre coordinating the movements,' he continued.

A black-clothed figure was pointing at a section of the crowd. His positioning and his stance both suggested he was in charge of events. It was impossible to tell what he was demanding of the other armed figures who were sending the crowd in the direction he was pointing. After another half minute or so a distinct crowd line became more obvious. As more stragglers joined in from the access roads, they were soon standing three or four deep.

The three crew members took different parts of the screen to focus their attention on, picking out more armed jihadists in the process. Whatever was happening had taken some serious coordination.

The splitting of a crew's focus in that way had become increasingly important over several years of Reaper operations.

It stopped all three sets of eyes from being drawn to the most obvious feature on the screen and missing something crucial on the periphery. In the right circumstances and with the right visual stimulus, the most highly trained and disciplined brains in the world can instinctively react like curious puppies. For a Reaper crew, during high-intensity events it became even more important to maximise the available collective brain power, using effective CRM. It is a concept that is considered to be so important that it merits its own TLA (three-letter acronym).

As the Reaper continued its orbit, Brodie finessed the camera controls to provide the most helpful views of events.

'Two armed men on the roof of the high building at the bottom of the screen. Confirming weapon type.' Ricky was now in touch with his intelligence counterparts in the CAOC, getting extra assistance with analysing events. The two men were standing at opposite ends of the building that overlooked the junction, both looking down on the crowd. From their vantage point the two snipers dominated the area below.

'Vehicle entering from the top right of the screen,' observed Brodie. A van edged its way through the crowd, which was being policed by rifle-wielding IS fighters. He shifted the camera angle to put the van in the centre of the screen.

'Roger,' confirmed Gav. He picked up one of his telephones to speak directly to the Auth in the Ops Room upstairs. 'Do you have eyes on our video feed?'

'Roger, we have eyes on. They also have eyes on at the CAOC.'

The RCH was a former XIII Squadron SO, and she was in direct communication with the Auth. She informed the Auth that the live video feed was now up on the big screen in the CAOC at Al Udeid Air Base, where she was now watching it. Everyone in the Command Centre could see what Gav and his crew were doing, including the Target Engagement Authority

Colonel in overall command of that day's air operations.

In the few seconds it took to let Gav know just how many people were observing the video feed from his Reaper, the prisoners were pulled from the van.

'This is an execution.' Gav's tone changed slightly as the adrenaline kicked in. The trio started working through the options. They immediately ruled out the possibility of using a 500lb guided bomb because of the crowd being herded in to watch.

'Shit.' The trainee pilot was still in the GCS with the crew. She did not have to see what might be about to happen, so Gav sent her out. Then the Ops Room was cleared of all non-essential personnel. There are some mental images and memories that are to be avoided where possible and an IS execution is one of them. By 2017, it had become common practice on the Reaper Force to reduce the number of people watching particularly gruesome events on the ground in Iraq and Syria.

The three crew members started to methodically work through the options, none of which was appealing. Any impact point near the thick crowd line would be disastrous. There was a slim possibility that a missile to the centre of the execution area *might* not hit people in the civilian crowd, but it would definitely hit the men who were about to be executed. There is no such thing as a legal mercy killing in the Law of Armed Conflict, so that was not a viable option.

Gav's leg twitched faster.

'There's no good option.' Frantic reworking of the options led to the same conclusion: no strike. They were going to watch, powerless, as the men below were publicly killed. It was probably an act of intimidation of the local population rather than the fulfilment of what would be considered justice in some parts of the world.

The crew started to give up on the possibility of preventing the inevitable. Heads dropped slightly. Gav and Brodie briefly glanced at one another. They were all looking and re-looking. From the squadron Ops Room to the CAOC thousands of miles away, several vastly experienced Reaper personnel were looking for a shot that wasn't there.

The telephone light flashed and the Auth picked up. A woman's voice came on the line from the Command Centre. It was the RCH.

'There might be a shot.' She had extensive experience as a Reaper SO and had an idea that might work. Risky, but possible. She also knew Gav and the rest of the crew involved, and knew they were good enough for a shot with almost no margin for error.

Her calmness contrasted with Gav's rising stress levels.

'The two snipers on the high rooftop. Strike the one on the left of the screen – the blast should disperse the crowd,' she suggested.

The Auth relayed the message to the crew.

'Roger, zooming in,' added Brodie.

'CDE?' asked Gav to his two crew colleagues. They started a collateral damage estimate and began looking for a 'shift cold' area.

Ricky got right on it. 'The snipers are looking over a four- to five-foot high concrete wall that runs round the top of the building. Hit the sniper on the left and the wall will stop the missile blast from hitting the crowd below. At worst, there might be some small bits of debris.' Brodie concurred.

'Worst case?' asked Gav.

'The Hellfire goes in slightly too high and over the top of the wall. It would hit the middle of the square. The prisoners will probably be hit along with their captors. If it goes even higher, civilians in the crowd would come into the blast zone.' Brodie was pensive but he could see the shot. He would have to put it on the target – the sniper on the left – with a margin of error of only a couple of feet.

Gav started bringing the aircraft round to get the correct angle of attack, at the same time requesting approval for the strike. He and Brodie had flown together many times and he trusted the SO implicitly. Another glance at each other. Slight nods.

Approval took seconds. They knew that the full team of advisors, lawyers and the RCH were watching and working in tandem.

'9-line approved for the sniper on the top left of the building.' The RCH gave her permission.

Unusually, Gav's heartrate was going down rather than up. It had been at maximum when he thought there was nothing they could do. Now, he just faced the pressure shot of his career. In some ways it looked easy – a static target. What made it so difficult was the impending human tragedy if it went wrong. Plus the fact that the world, or at least a huge military infrastructure, was watching events develop across multiple time zones.

Some of the best golfers in the world have missed easy putts in the cauldron of the Ryder Cup. The pressure disrupts the signals from the brain to the hands. Brodie's whole body was now surging with adrenaline as he prepared to guide the missile onto its target.

In the square below, the prisoners were being hauled into position. From the gesticulations of the dominant black-clad figure near them, some form of speech or proclamation was being made. Was there an IS video camera somewhere in the vicinity ready to produce the latest shock images for social media?

'Cleared hot.' The voice on the other end of the radio sounded calm.

'Final checks. Everybody content?' Gav double-checked the missile choice and setting as Brodie and Ricky confirmed that they were ready.

A message came through on Gav's secure internet link to the CAOC.

'Do it now.' Events had developed so quickly, with airspace needing to be rapidly reorganised, that the colonel in overall command gave the final instruction.

'Three… two… one… Rifle.' The slight wobble of the camera image an instant later confirmed that the Hellfire had blasted from its rails towards the figure with the rifle.

Brodie was trying to keep a relaxed grip. He would later recall how his heart was pounding so hard it felt like it was trying to jump out of his chest, while at the same time his legs felt 'fizzy'. Any sudden jerkiness of the crosshairs would be catastrophic. Small, gentle movements. The white cross shifted fractionally but stayed on the tight target area.

'Fifteen seconds.' Gav counted down to impact as he and Ricky scanned for anyone coming onto the roof in the vicinity of the sniper they had in their sights.

No movement.

'Ten seconds.'

Two seconds later the crosshair danced fractionally off target.

'Fuck,' muttered Brodie under his breath. He suppressed a spike in his heart rate and made the gentlest correction. It had to be right – there was no time for any more changes.

As Gav started his final countdown, the satellite time delay to the missile guidance had passed and the crosshairs were edging into the middle of the target area.

'Five…four…three… two… one… Splash.'

At some point during the last instants before impact the crowd reacted. They most likely heard the incoming missile. When the footage was subsequently replayed and reviewed the movement could be seen clearly. It can also be seen on the video segment later released by the MoD.

The jihadist sniper at the centre of the screen was instantaneously replaced by the missile blast and a cloud of dust. In the GCS, in the Ops Room next door and in the Command Centre thousands of miles away, eyes strained as the cloud cleared. The unspoken question was the same everywhere. Had the plan worked?

Brodie panned out slightly to give a wider view of what was happening below. The crowd was running in every direction. Whatever control the IS fighters had been exerting was trumped by an instant of collective instinct to flee an even greater danger.

Ricky was concentrating on the spot where the prisoners were being held, ready to be killed. He rewound and replayed the first few seconds of footage to see what happened to them. They had managed to merge into the fleeing mass of people, and it looked like they had escaped their IS handlers. Then they disappeared from the view of the camera.

As the blast cloud cleared, there was no trace of the sniper who had been hit.

A quiet 'good shot' was all that passed between Gav and Brodie, echoed from the rear seat by Ricky, and acknowledged with a nod. The sentiment was echoed from the CAOC. This was a moment for relief, not celebration. The crew also knew that without the suggestion from the RCH watching from the Command Centre they would most likely have witnessed two beheadings. They might also have had to watch while the executioner beckoned young boys to pick up and pose with the heads for IS social media.

They had to put aside a whole series of random thoughts, especially the '*What if?*' question that kept coming back. There was much still to do, starting with BDA.

There was damage to the roof and to the short wall that the sniper had been standing behind. Crucially, the top corner of the building where the sniper had been standing was still structurally intact. If the building had collapsed – or even if a large chunk of it had fallen – civilians below would have been crushed.

In the immediate aftermath, the emotion and adrenaline began to slowly subside. If anything, hands trembled slightly more, partly because the greatest moment of danger to the civilians had passed and partly because of the mental distraction of the other post-strike activities. Gav still had an aircraft to fly and he had started a slow orbit of the area. The missile had been fired from the direction of the rear of the building towards the execution square in the middle of the crowd. In order to check for potential casualties on the ground in front of the building the camera would need a clear line-of-sight.

Ricky was carrying out multiple intelligence-related tasks simultaneously. As well as keeping one eye on the live video footage, he was re-watching crucial elements of the recordings of the past minute or so. In between, he was updating the online chatrooms to the centralised intelligence sources he relied on and contributed to. Today had been a good day for the intelligence community. They might not have known exactly what was due to happen, or precisely where, but they had known enough to get the Reaper in roughly the right place at roughly the right time. Sometimes that is as good as it gets. But nobody overlooked the element of good fortune.

The initial scan of the front of the building for prone bodies was positive – there were no casualties. With that confirmation most of the remaining tension ebbed away. Most. They would also have to face the debriefing officer. Regardless of the tension involved and importance of getting the strike right, it would still be debriefed like every other Reaper weapon event. And the debriefing officer would not be blowing sunshine up their collective backsides just because it turned out well. His job was to ensure that Gav and Brodie were performing at a level to justify being elevated to QWI. It wasn't just about getting a weapon on target. That was expected of everyone. It was about taking the initiative in an incredibly tight and challenging space, making sure that key decisions to act – or not act – were made at the right time, and using all of the resources available to them.

Time, as always, took on an elastic texture. Before they knew it, Gav, Brodie and Ricky were being replaced by the Op 2 crew so that they could take a break, catch breath and start to reflect on what they had just done. As they walked back to the Ops Room upstairs, they knew they would have to refocus, get back to planning their QWI course mission for later and find out what the rest of the shift had in store for them. They had been away from the planning room for just ten minutes.

This missile strike has the distinction of being one of the few whose video footage has been released to the public by the MoD, in this case some months after it took place. The week it occurred I happened to get in touch with Gav to do some follow-up with him and his wife Penny about an event covered in another chapter. He mentioned that he had been involved in an incident that I might be interested in. He would ask permission to show me the full footage if I came to XIII Squadron and also talk me through what had happened. He could not provide any more detail over an unsecure phone line.

Intriguing. When a Reaper crew member describes something as 'interesting' I had come to understand that it certainly will be. Members of the Reaper Force have, to different degrees, a distorted view of the world and what passes for 'interesting'. From an outsider's perspective everything they do is interesting because, as operational military personnel, their job operates at the extreme of human behaviour, i.e. killing people in war. They also have to watch atrocities being carried out, from IED deaths to IS executions. Mentally, they go where the rest of us don't. And this can have a number of effects over time. It can change how individuals view the world and how they interact with others outside the GCS.

The regular occurrence of extreme experiences over time can lead to what one former Reaper Squadron Commander described to me as a kind of parallel 'normality'. Some of what counts as 'normal' on a Reaper squadron would not be considered 'normal' anywhere else. Not in the sense of wrongdoing, because the operators see the use of weapons and lethal force as a means of acting for good against enemies who commit terrible acts, often against civilians. It is more of a sense of moving in and out of an alternative reality that almost looks familiar but really isn't when you scratch beneath the surface. Something that is difficult to fully appreciate from the outside.

My first day with XIII Squadron at RAF Waddington brought home to me the co-existence of two 'normalities' that the crew members move between. In some ways, the difference seems starker at XIII Squadron in England than at 39 Squadron based outside Las Vegas. The latter already has a surreal quality to it, making it feel like an abnormal place for British personnel from the RAF, Royal Navy, Royal Marines and British Army to be working. But back to the day I met Gav.

I arrived at XIII Squadron just before lunch and was taken into the crew room for coffee. I was faced by a scrum. Backsides covered with green flying suits pointed in my direction as their owners grappled for something in the fridge. Allegations of theft abounded, along with claims of non-entitlement and minor threats of violence that I took to be jokey but which might not have been. The day's rations had just been delivered and a group of around eight grown men and women discovered their inner child. Their inner, greedy, chocolate-loving child. They were fighting over Twix chocolate finger bars.

'No duty, no Twix. They are only for personnel on the flying rota.' The person in charge of distributing the rations was losing the argument. Actually, they were being ignored. I always like to see good equal opportunities in practice and here was a great example.

As the rations officer retreated in defeat, a small cheer went up and the Twix-wielding scrummagers emerged with their trophies. Having fought the good fight over 60 pence worth of chocolate fingers, three of the group sat together at the breakfast bar next to where I was being served my coffee. They were obviously a duty crew on a break and they resumed a conversation that must have been interrupted by the rations officer.

'We won't have to worry about the blast effect hitting the buildings if we run in from the east,' proposed the pilot.

'Yes, but if the Hellfire comes in low it might just clip one of the buildings.' Mugs were set out to represent the physical layout of buildings they had just been watching in Syria. Chocolate fingers were being used as dummy missiles to demonstrate the strengths and limitations of different angles of attack from different directions. They were calmly discussing the best way to target and kill someone. Using Twix.

Parallel normality. And it works just fine when people mentally constrain it to their work. Ideally, they will switch off when they get home and tune into the 'normal normality' that the rest of the world lives with. However, few seem able to do that consistently and what they do at work seeps into their home lives. For some – and I am still not able to gauge how many – that seepage can eventually progress into degrees of mental trauma. For a small number, that mental trauma can be severe. Given my reaction to the first time I saw people killed, I suspect I would be one of the more affected ones, but I will never know.

But returning to the question of medals and that strike. The missile killed the sniper, disrupted the execution and dispersed the crowd, preventing physically coerced Syrian men, women and children from having to watch the beheadings. Also think about the Reaper crew at that moment. It was a very difficult shot with the added pressure of it being watched by a large number of people in the CAOC – dozens at least. If the SO had been inaccurate by a yard, with a missile travelling at hundreds of miles per hour, buffeted by wind and controlled via satellite with a time delay from his control stick, they would have killed and maimed many civilians. Why take that shot? Why risk spending the rest of your life with the mental burden of being the person who killed multiple innocents instead of stopping an execution?

It takes a different kind of moral courage to risk your mental wellbeing for a distant stranger in the line of duty. Such actions will never be seen as heroism in the conventional military sense because of the lack of physical risk attached. Yet it will become an increasing element of military operations in the future: on land, at sea and in the air. This gives the UK government – and any other that uses remotely operated aircraft like the Reaper – an interesting dilemma. On the one hand, government policy for the NHS says, 'We must provide equal status to mental and physical health.'[31] Yet when it comes to medals for a new kind of war in the twenty-first century, a nineteenth-century approach that only recognises physical risk and physical harm is still used.

31 Independent Mental Health Taskforce to the NHS in England, 'The Five Year Forward View for Mental Health', p. 5, https://www.england.nhs.uk/wp-content/uploads/2016/02/Mental-Health-Taskforce-FYFV-final.pdf, accessed 7 May 2017.

I should declare a personal interest in the long-term psychological wellbeing of military personnel. Between July 2004 and July 2005 I was an RAF chaplain based in the Falkland Islands. During that time I conducted twenty-four private memorial services – over and above national remembrances – so that returning veterans of the 1982 conflict could remember and pay respects to their colleagues who had died in the war. Most of those veterans who returned to visit the Falkland Islands during my year there had been suffering from some form of mental trauma: sometimes in addition to physical damage, sometimes not. The majority of those I met had taken between *ten and twenty years* to recognise something was wrong with them and seek help.[32]

Over the subsequent couple of weeks after the medals announcement, it became clear to me how demoralising it was for many members of the Reaper community to have their efforts publicly discounted when it came to their contribution to Operation SHADER and the fight against IS. It seemed not so much about medals than about recognition. Some of the spouses and partners took it worse than the crew members themselves. Perhaps that is a result of seeing your loved one changing before your eyes over time because of what they do, and having their dedication dismissed as somehow unworthy of recognition.

So it seems like a watershed moment for official recognition of Reaper personnel and what they do. Or, at least, recognition of some of the major calls that they make and the possible lifelong psychological repercussions if things go wrong. Chapter 5 highlighted the consequences for someone in the crew that killed four civilians in 2011. He carries that burden and always will. That is the nightmare scenario for every person who flies a Reaper.

32 A BBC report on the mental health of Falklands veterans can be found at http://www.bbc.co.uk/news/uk-22523317, accessed 20 January 2018.

Unfortunately for the Reaper crews, history, tradition and ignorance are all lined up against them when it comes to the matter of medals. The eminent RAF historian, Peter Gray, has observed that in the Second World War the RAF lost out when it came to medals and official recognition because the Committee on the Grant of Honours, Decorations and Medals in Time of War did not understand how air power had changed war.[33] In 1945, the RAF Air Member for Personnel wrote the following to the Permanent Under-Secretary, the Chief of the Air Staff and the Secretary of State:

> The [Medals] Committee gave the Air Ministry representatives a sympathetic hearing. But the fact of the matter is that the Committee is comprised of senior civil servants, generals and admirals who are really incapable of thinking of war except in terms of battlefields and 'theatres of war' geographically defined, and of course everything afloat. They cannot understand that the air has changed everything.[34]

If the final sentence said, 'They cannot understand that the Reaper has changed everything', it could have been written in 2017. *Plus ça change.*

Which leaves me with a question. Would I take on that potential nightmare of killing civilians if my efforts were maligned in many quarters and not officially recognised by the government that sent me to do the job?

33 Peter W. Gray, 'A Culture of Official Squeamishness? The Air Ministry and the Strategic Air Offensive against Germany', *Journal of Military History*, Vol. 77, No.3, July 2013.

34 AIR 3/9303, Minute AMP to PUS, CAS and SofS dated 31 August 1945, cited in Gray, 'A Culture of Official Squeamishness'.

POSTSCRIPT

On 18 July 2018, the UK Defence Secretary announced that the Operational Service Medal Iraq and Syria, the Operation Shader medal, 'will now recognise those making a vital contribution to Op Shader from outside the conventional area of Operations, for example those Reaper pilots taking life and death decisions from back here in the UK.'[35] Time will tell if this offer comes with a host of caveats or whether there has been a genuine shift in Government attitude to the contribution made by the Reaper Force to operations in Iraq and Syria.

35 Ministry of Defence, 18 July 2018, https://www.gov.uk/government/news/new-medal-unveiled-to-recognise-the-fight-against-daesh, accessed 20 July 2018.

CHAPTER 12

MARRIED TO THE REAPER

'FROM LAS VEGAS TO SKEG VEGAS.'

LUCY, SENSOR OPERATOR'S WIFE

According to Greek mythology, the gods of Mount Olympus oversaw every aspect of the lives of humans. Through Hades, they even controlled the world of the dead. Though invisible to mere mortals, the presence of the gods was always felt, shaping and controlling every aspect of people's lives. The gods could be benevolent and passionate, but were also capricious and cruel.

It is generally assumed that the age of the Greek gods has passed; the temples are crumbling and their presence no longer felt. But anyone who has been married to a member of the RAF, or to a member of the other armed forces, will testify that a trace of their presence remains. They will tell you that the fickle finger of fate, which exerts unseen control over lives and destinies, is alive and well, embodied in the 'posting gods'. More correctly known as Desk Officers, these are RAF officers – usually from the same branch as the people whose careers they oversee (aircrew oversee aircrew, and so on) – whose role is to manage careers and ensure that positions in the RAF are filled by the right people at the right time.

To anyone who gets a series of great jobs in ideal locations, the posting gods are worshipped. However, to those who married someone in the RAF, the posting gods sometimes seem determined to make life as miserable as possible for them and their families, through frequent or unwelcome moves across countries or continents. Sometimes they even manage to inflict frequent *and* unwelcome moves, while every-so-often producing something good, and keeping cruel hope alive. Civilians who have only lived in two or three places might not appreciate what it is like to cede so much control of their lives. But they can probably imagine how they would feel about introducing their child to an eighth school in eight years (as happened to my daughters).

The Desk Officers not only recruit members of the RAF, Army, Navy and Marines to the Reaper Force. They also help to decide whether that new Reaper recruit – plus family, if they have one – moves to XIII Squadron at RAF Waddington and a house in Lincolnshire or to 39 Squadron at Creech Air Force Base and a house in the Las Vegas suburbs.

This chapter continues by looking at how different couples found themselves on the Reaper Force, before going on to explore different aspects of family life. This is where Reaper families differ from just about everyone else in the Armed Forces, because they send their husbands, wives, partners and parents off to war every day. They also get those same husbands, wives, partners and parents *home* from war every day, for years on end. When I say 'same', however, I don't actually mean 'same' because how can you remain the same when you might have taken lives, or watched IS murders, from afar, mere minutes or an hour or two before a family dinner?

INTO THE REAPER FORCE (1) – LUCY

I had one long-held ambition from the time I did my degree in Modern Languages: to live and work abroad, at least for a few years. I held on to that ambition through the years when I became a Modern Languages teacher and my career began to develop. So the possibility of foreign travel was a definite plus when I met, then later married, Ed.

Ed was an RAF fast jet navigator when we first got together. He was a bit of a high flyer – like Goose from *Top Gun*. That was a whole thing: marrying into the Air Force. It was part of his appeal – he fitted into what I wanted to do long-term. As well as being a great guy.

But the overseas bit never happened. We never ended up anywhere exciting, just the shires. We were like the Hobbit, going from Oxfordshire to Buckinghamshire to Hampshire. Despite that moving around, things went well for me, career-wise. I ended up in a primary school teaching job that I loved.

Ed had a setback with a medical downgrading that meant he could not fly in fast jets any more. So he started looking for other flying possibilities that would allow him the same kind of job satisfaction. Flying the Reaper as an SO became a possibility: cutting edge, the way of the future, and a real, front-line challenge. But I would have to leave my dream teaching job.

Then one day Ed arrived home, waving an official piece of paper and a bottle of champagne…

'We're going to Vegas, baby.'

'What?' I replied.

'I haven't told you but I've known about it for ages. I wanted to wait until I got the assignment order. It's here.

Vegas. I am joining 39 Squadron at Creech Air Force Base in Nevada.'

Cue delirious celebrations and plans. I would no longer be applying for the perfect deputy head teacher position. I was going to fulfil a lifelong dream instead. America here I come...

So I resigned from the school and sold my car. Who needs a right-hand drive car in Nevada? My brother planned a Christmas trip to come and see us. And then... and then... The Desk Officers stepped in. Ed arrived back at home one day: no champagne, no celebration and a new assignment.

'Change of plan. I've been dreading telling you about this. I've known for a few days – I knew you'd be upset. I've been reassigned to XIII Squadron at RAF Waddington in Lincolnshire.'

Experienced officers with Ed's background and skills were needed to help set up the new Reaper squadron at Waddington. Another 'shire'. Lincoln*shire*. I just thought: *Oh my god. Las Vegas to Skeg Vegas.* I didn't even try to hide my disappointment. I couldn't. ['Skeg Vegas' is an ironic nickname for the famously bracing Lincolnshire coastal resort of Skegness. It has many redeeming features when compared to Las Vegas, but it's an acquired taste for those who are not from there.]

And so began our journey into the Reaper Force.

INTO THE REAPER FORCE (2) – BRETT

I knew from the look on Zara's face that she had made up her mind about going to the Reaper Force. Kinloss had been our home for twelve years. We both worked there, we both had friends there. We both went through some bad things in 2006 when the Nimrod crashed in Afghanistan.[36] She was due to fly that day from the same location. For me, back in the UK, it was hard. Especially the fact that for six hours that day I'd thought she was dead.

I knew that that Saturday she should have been flying. I was away with the cadets when I got a phone call from a friend at RAF Lyneham asking, 'Where's Zara at the moment?'

'She's away, overseas.'

'Whereabouts?'

So I said where she was.

'Have you seen the news?'

No, I'm in the middle of a field with a bunch of teenagers. What's in the news?

'A large aircraft has crashed.'

'OK.'

At that time I didn't even think about the Nimrod. It could have been a C-130, anything. Then she said, 'Oh, it's just come up on the news: it's a Nimrod.'

So at that point I thought… *Right. Zara will be flying.*

So there's me in the middle of a field trying to put the radio on. There was very little news at the time. I was sitting there with some of my friends and I didn't know what to do. I got the number for an MoD phone helpline but it didn't give any details. I didn't know what to do. Jump in my car and go to Kinloss to get the news right away? But my friends took my car keys from me. They didn't think I was in a fit state to drive.

36 On 2 September 2006, RAF Nimrod maritime reconnaissance aircraft XV230 crashed in Kandahar Province, Afghanistan, killing its 14 crew members. The report of the Independent Review by Charles Haddon-Cave QC, of the circumstances surrounding the crash, can be found at https://www.gov.uk/government/uploads/system/uploads/attachment_data/file/229037/1025.pdf, accessed 20 February 2018.

A few hours later I got my keys back and I was sitting in my car ready to go and my phone rang. It was Zara. There had been a comms blackout. She told me she was safe. But people who I knew, people who I'd worked with, were on that aircraft. Although I was relieved that she was alive, it was upsetting that people who I knew had died.

The important thing back then was the support for the families who had lost loved ones.

So a few years later when we started looking at Zara joining the Reaper Force, that was one factor – the crash – we wouldn't have to worry about.

INTO THE REAPER FORCE (3) – IONA

My partner, for his own reasons, had, behind my back – he tends to do these things – phoned his Desk Officer at Manning and was offered a place on the Reaper Force. That wasn't anything that we'd discussed. And I can remember being in the en-suite upstairs and him saying, 'Oh…' – and this is very much my husband's style – 'I've got to speak to you about something…' And he came in. I was holding my son, who was about eight months old at the time.

'I've been speaking to Manning and I think I want to go onto a different aircraft type.'

I said, 'OK…'

'It's the Reaper.'

I didn't know anything about Reaper.

And he added, 'It's in America.'

At which point I think I actually did say to him, 'Are you having a joke here, are you pulling my leg?' I thought, *I really don't know what to say to you.*

'We have seven days to make a decision about whether

we are going or not.' I was on maternity leave at the time.

So we hummed and hawed. I think because of all that had happened when we had been around the Nimrod crash, I thought, *We need a bit of a break.* So within seven days we had decided. We said, yes, we'd go for it. Life's too short not to do it. But I didn't know anything, really, about the Reaper.

But I did know that he wouldn't be inside it. I thought: *Great, he's going to be in an aircraft type that's not going to blow up. He's not going to be in it.*

INTO THE REAPER FORCE (4) – EMILY

He went into the flying training system but it took him a long time though. He was 'holding' – waiting between the different phases of flying training – in various posts, and it did seem to take for ever. He went to RAF Cranwell for Elementary Flying Training and then to RAF Linton-on-Ouse for Basic Fast Jet Training. It was towards the end of his flying training there that the opportunity came up to join the Reaper Force. It was a question of, you either volunteer or you're going to get told if you're going to do it or not.

We talked about it lots. We decided that Reaper would be a good way to go for his career because it was taking so long to get to through flying training, and he was already in his mid-twenties. That doesn't seem old, but when you have to get promoted to a certain level by a certain age then actually it's quite significant. That's the main reason we went down the Reaper route, really. That, and we thought we'd see each other a bit more.

FAMILY LIFE (1) – LUCY

Becoming operational aircrew on the Reaper involved Ed doing his SO training in America, followed by some to-ing and fro-ing between 39 Squadron in Nevada and XIII Squadron at RAF Waddington as he built up essential experience and qualifications. As a Flight Commander, he would also have additional leadership and management responsibilities on XIII Squadron, as well as the six days on and three days off shift pattern. It meant that in the early months of his time on the Reaper Force, he was not around as much as I would have expected – or liked.

It didn't help that I had managed to get a teaching job. I was working full time, and Ed was doing his shifts. We'd not see each other for weeks and weeks. I remember I thought, *This bedding's not been changed for weeks*, because if one of us wasn't in it the other one was. *This bedding's walking to the wash – we need to take it off.* Sometimes, all we'd say for several days in a row was, 'Morning love, see you later… Night, love.' It was pretty rubbish.

In the middle of it all, I had what felt like a bit of a mid-life crisis. *I've not seen my husband for ages and I've given up my fabulous job. I've now got this rubbish job where I've just handed my notice in, and I'm on my own. I've done exactly what my dad said not to: I've given up my house, career and friends to follow someone around the country when he's not even in the country – he's in a different country. What am I going to do with my life?*

I was thirty-eight at the time and thought, *Bingo, maybe I need a child.* I am sure I was joking as I thought it. But then, thoughts like that – even joking ones – can help shape a person's life plan. How hard could it be?

Once we overcame the obvious problem of two people needing to be in the same place at the same time to start a family, I found myself pregnant. Then Ed came home with a pronouncement: 'I'm being deployed to the Middle East for six months.' He was to travel out to Al-Udeid the day after my due date. The Desk Officer let Ed delay his departure a bit. So the baby was nine weeks old, screaming his head off with a dairy allergy, and he was off to the sunshine for six months.

FAMILY LIFE (2) – BRETT

When Zara was getting ready to go over to the States to do her training at Holloman Air Force Base in January 2014, our daughter was three at the time and she didn't know what was going on. I was left to pack the house up and move down to Lincolnshire in March. I had no contact with the squadron at that stage.

In April 2014 I went out to the States and Zara showed me a Reaper – I was quite surprised by the size of it, with its sixty-foot wing span. That's big. She showed me a bit of the base – what she could show me – so I arrived back here with a greater understanding. My mum looked after our daughter for a week. It sounds bad but it was a bit of a break from the move, from setting up this place, then getting our daughter into a playgroup. And of course, back then I still didn't have a job. I was effectively a single parent at that time. I had no family nearby.

Zara got back from the States in June. I think she got back on a Thursday or Friday and started at XIII Squadron on the Monday. Then she came home and explained about the six days on and three days off shift pattern… We wouldn't get every weekend together, but we would make the most of the three days off and make it work.

Then in July she went onto permanent nightshift for twelve weeks, so we didn't see each other. She would leave at six in the evening, return at seven in the morning and sleep. Get up, get something to eat and go to work. I felt quite lonely.

TRANSITIONS – APRIL

It is a really weird situation. I often think about it: how weird it is that he does what he does at work, and then coming home and being a loving and caring dad and husband. That transition between the two is a really strange one, and one that nobody other than Rory can understand in this context. I obviously have no idea. He can't tell me everything that's going on, that would be wrong. But they need other people to offload to.

Rory is immersed in war imagery *all the time*. That's got to have an adverse effect at some point. It has to. I think that a lot of people find that it does have an effect. Rory particularly, has done a really long stint now – five years. Which, for war, is a long time. It is such a long time to leave somebody in that position.

He does get frustrated and has a shorter temper now. But I think it's just an avenue to get rid of whatever it is that gets created at work. And home is the safest place to do it.

ON THE DIFFERENCE BETWEEN BRITAIN AND AMERICA – IONA

If the Americans realised you were in the military, you would go for your veteran's buffet lunch, which you and your partner would get for free. There was always something. One free trip a year, with up to two or three children, to Disneyland. Although we were British military we worked with the USAF, so we got all these perks. *Everything* got supplied. One of my friends who was out there was fumbling to get money out when she caught a bus. The lady behind realised – she'd seen the military card – and said, 'Sit down, sit down,' and she paid. It was her way of showing support for the military.

Since I came back to the UK, a big thing for me is having to hide what my partner does. Back here you get a protest that happens once a month, and it does hurt. As you go past them you see the signs: 'Drone Killers', 'Kill Babies,' and all this kind of stuff.

Even at work, people will ask, 'What does your husband do?' And I kind of have to stop myself and say, 'Oh, he does the AWACS[37] thing. I don't have much to do with what he does. It doesn't interest me.'

And I think: *Why am I lying to people? This is ridiculous. I'm a grown woman with children and I'm lying.* But it's because I wouldn't feel secure with that information going out. Obviously, with close friends it's fine. But we're told not to go overboard with what we tell people especially here in the UK and around Waddington.

From an American point of view – not that I am saying we should run our country like the Americans – it's so nice to be received positively. You'll go into shop and get military discount. Every shop you go in. I can remember going to Newcastle with one of my friends who was also

37 'AWACS' stands for 'Airborne Warning and Control System'. The aircraft is the E-3D Sentry AEW1. It is based on a Boeing 707 airframe and is best known for the huge rotating radar disc that it carries above its fuselage.

based in Vegas. She went into a shop to buy a handbag, and she asked, 'Do you do military discount?'

The woman could have cut her down with a dagger: 'No, of course we don't. We just do student discount.' Things like that make you realise you are treated *so* differently in the UK. It's quite difficult to come back to that.

WHAT HAPPENS AT WORK (1) – LUCY

I'd ask Ed, 'So, did you kill any baddies today?'

And he'd reply, 'I couldn't possibly say.'

I'd say, 'Come on, you do this every time.' I was just interested in knowing about his work, but it's very secretive.

I know that a lot of the time Ed's job was just watching people on the ground, but occasionally it would involve firing a weapon. I would have loved to know more detail. He would be *so* professional and not say a thing. Whereas with his previous job, we used to talk about that all the time. Of his Reaper work he says, 'I couldn't possibly say. I could tell you but I'd have to kill you.' And he'd smile.

I don't know if it's his professionalism or if it's just his personality. He'd just switch off completely at the end of a shift. He never came home upset. He'd come back a bit stressed because he was busy but that was the nature of his leadership and management role as a Flight Commander rather than his operational work. But the crux … the purpose of the Reaper… their missions… he wouldn't talk about them at all. Or there'd be something on the news and he'd say, 'Yep, I knew about that for weeks. I know who pressed the button as well.' And that was all I'd get.

WHAT HAPPENS AT WORK (2) – ZARA

It's so hard on the family, what we do. It's so difficult for them. Because you live it, you breathe it, you sleep it. You dream it at night. You wake up in the middle of the night chewing your cheeks, grinding your teeth, just thinking about it – what you could have done better.

All of that has an impact on your personality, how you come across to your family. After my Reaper flying tour, I was quietly pleased when the Boss offered me a ground tour. I've been shattered, my family have been shattered. But it was also the most satisfying thing I have ever done, especially anything that involves self-defence and stops attacks on friendly militaries.

WHAT HAPPENS AT WORK (3) – LISA

I do think it takes a certain type of person to deal with what they do. I think he's conceptualised that into boxes: 'This is my job. This is what I'm paid to do. I can justify why I'm doing it.'

For me, I'm proud of what he does. Immensely proud. I suppose, and I know this sounds silly, but my husband is the person who comes in the door and changes nappies, walks the dogs, puts the kids to bed and does everything that a father and husband would do.

We were at the mess and one of my friends said to me, 'Do you know what, if you look at our husbands, you could actually say that they're killers.'

And I replied, 'You can't say that.' She wasn't saying it to be bad, she was just being a bit flippant.

'What other way do you want to put it?'

It was just this bizarre conversation over wine, and I'd never looked at it that way before. I was shocked that somebody could think that. So I did a bit of research on my own, to understand the difference between what military people do and what killers do. I had to justify it all again in my head, to get to a point where I was comfortable. I do think there are a few military people who struggle with those killing elements. But I think that's good. Because it's the death of someone. It's a major, major thing. They're not delivering Tesco's food then coming back to their wives. They are doing these major things.

WORRYING – LISA

A lot of the time I would be in bed thinking, *He was meant to be home at nine o'clock and it's now three in the morning. And I haven't heard a thing.* I'd try and get back to sleep, then I'd try his phone. I used to panic because when it got to Day Five or Six of the six days on shift you'd see the blood-shot eyes from the fatigue.

It wasn't unheard of in those early days at Creech for him to say, 'I had to pull over to the side of the road and sleep in the car for an hour on the way home because I was nodding off. I nodded off three times.' On two occasions he told me that he'd been woken up because he banged the kerb on the side of the road because of the level of exhaustion he was experiencing.

We were out at Creech for just short of three years and that tempo was continuous. At points I would be ringing and ringing. You know when you get something in your head. I'd think, *He's not back, he's not back.* Then I'd rationalise, *It's OK, it's OK, he'll be fine.* I'd try to get back to sleep but I couldn't. Not until he got in.

I'd start to think, *I don't want to phone the squadron – I'll sound like a moaning wife. But if he left three hours ago and he's still not home…*

I'd sit online and look at the police reports for any crashes on his route home. At the worst point, for five out of six shifts he would be at least three hours later than he said. The fatigue levels on him must have been astronomical. But, when I look back on it, the fatigue levels on the family with the cycle of work was crazy as well, actually.

But the type of character my husband is, he thrives on that. When we were at university, if he had a dissertation due in, he would be up till three in the morning with Red Bull, and his eyes would be bleeding. But that's how he is.

Now, in 2016, the shifts have improved remarkably for the better.[38] We had another two children in those three years in America, so it couldn't have been terrible.

What made it bearable was that I could see he really enjoyed what he was doing. He had a real passion for it. He wasn't doing it to stay away from me and the children. It wasn't that he was trying to stress me out with all these things. He felt that that was his job and he had to do it to the best of his capabilities.

From a mental point of view, he has always seemed to be coping with what he's doing. I did initially worry about that. How can you not be affected? We had chats about that, and he seemed to be fine. But the physical impact on his body, lack of sleep, constant mental exhaustion – just not getting enough 'down time'. That's what I worried about most. But I

---

38 Note: This was said at RAF Waddington. And because of the eight-hour time difference between Waddington and Creech, if one of the squadrons is working family-friendly hours, then the other probably is not.

don't think he thought about doing anything else – that was just his job. He was just a bit of a glutton for punishment.

SHOPPING – EILEEN

A few days ago my husband and I went to the local supermarket to stock up on food and supplies for the next week or so. As we were picking up some fresh vegetables, just inside the store, out of the corner of my eye I noticed him stop, shudder slightly, then carry on with the shopping. It was a bit like a robot rebooting in a sci-fi film, or seeing someone get a drip of cold rain down the back of their neck.

I asked him, 'Are you OK?'

He replied, 'Yes, fine.'

I wasn't convinced – I know him too well. So I asked again, 'What just happened? Tell me.' I was quite insistent.

He looked around to make sure nobody could hear, and he came clean. 'When I looked round this section of the shop, for an instant I realised it contains about the same number of people I killed in a strike yesterday.'

And then he kept pushing the trolley.

WORK-LIFE BALANCE (1) – SIMMO

I am a parent governor for my local school and every year I volunteer to go away with the teaching staff and help the kids enjoy the great outdoors. It's only three days away but the kids get to abseil, canoe, pot-hole and do many other fun things. One year, I had a great time and thoroughly enjoyed the company of the children and the staff.

Eighteen hours after I got back I was in work, watching a prisoner having his head cut off and being powerless to do anything about it. Oh how my life had changed – and not for the better – in such a short period of time.

WORK-LIFE BALANCE (2) – IONA

A lot of people who are married to someone on the Reaper Force can't do paid work. The shift patterns don't support it. I'm very lucky, I can dictate my own hours.

My eldest is seven years old and over much of that time I have felt like a single mother. Sometimes my partner is at work for fifteen or sixteen hours a day. So over six days the kids might see him once. And that's it. The sacrifices are significant.

I would say it's made me more independent. I was out of work for four years while I had the three kids, and I was a bit dubious about going back to work. But I thought, *My partner dumped me on my own in a foreign country for four months. I set up the utilities, I cracked on with it. I learned to drive on the other side of the road.* So from my point of view, from a confidence point of view, I did all these things. I brought up the kids.

I went into Nellis Hospital to have my third child. That morning I could feel some twinges. My husband's very words to me were, and this probably reflects just how committed he is to the job: 'Well, you better be sure because I've got my STANEVAL [Standards Evaluation quality control check on individual or squadron performance] check today and I'm not cancelling it if the baby's not coming. This is serious, you know, if I have to cancel it.'

I thought: *This is your daughter coming...* It's funny now but it wasn't at the time. He wasn't amused that he had to change his STANEVAL check.

So from the point of view of just getting on with stuff, I know I can do just about anything.

Some of our daughter's early words were, 'Daddy go work'. And she said that for about six months. 'Daddy go work, daddy go work,' because he was never there. I feel for him a bit because he's missed out on his children.

WHAT DOES DADDY DO? – LISA

From a family point of view there are difficult issues to deal with when they come home every day from the kind of work they do on the Reaper Force. For example, we go to the mess for a families event every month and there are pictures of all the different aircraft types on the wall.

My boys will point to the picture of a Reaper and say, 'That's my Daddy's plane.' The picture of it has the massive weapons underneath. And that's the first thing they spot. Then they say: 'That's Daddy's plane, he must blow people up.' They're seven and five, and at an age where I don't want to lie to them. We're teaching them right from wrong, and what's good and what's bad – what they should and shouldn't be doing. From a parental point of view I am really proud of what my husband does, but I'm not going to lie to my children either.

There was a Families Day when we first arrived back here at RAF Waddington a few years ago. I can remember that the Squadron Commander at the time had arranged to show the video of a strike, so we could see what our husbands did. He said in advance that it was going to be happening. The picture was fairly fuzzy – my kids would not have known what was going on. If anything, they would have thought it was a terrible cartoon. One of the other mums took her older children out because she didn't want them to see it.

I do explain to my sons that Daddy does go out to look for bad men, who have done terrible, terrible things. I tell them that there's good and bad in the world. I do it at a very basic level. A time will come when they will say, 'Does Daddy take people's lives?' That's for my husband and I to speak about, and to give an answer that is acceptable to them. But it's a difficult thing.

SOCIAL LIFE – CERYS

When we were on a Tornado squadron, the Boss would organise social events. The Squadron Executive Officers' wives would also organise events, and we'd all get together. It has been the same with other flying squadrons that we've been around.

Yet I've known some of the other wives on our current Reaper squadron for years – lovely people. Do we socialise with them? No, we don't. Because we can't. If the six days on-three days off shift patterns don't match up, then you just can't. Whereas on other squadrons we've been on, everyone has a much more regular, similar lifestyle that supported more communal activities.

PROUD (1) – APRIL

I am completely in awe of Rory and his pilot. I find it simply amazing that they were able to do what they did and literally save people's lives. There are not many people in this world that can honestly say they've saved a person's life and I am immensely proud to know, if in secret, that

that's my husband who guided the life-saving shot that people saw on the news. [See Chapter 11]

Unfortunately there are people in this world who do not agree with what my husband does, but I would like to say to them that my husband and his colleagues are not 'killers' they are protectors. They do a job that not many people could handle, they do it with a level of skill and professionalism that is beyond the norm and they do it for their country. Because they do what they do we can play with our children in the park and tuck them up safely in bed at night without fear; they are ridding the world of an evil that is incomprehensible. And I believe that if it came knocking at your door you'd pray that my husband would be there to save you from that evil. How lucky I am to be married to a true hero.

PROUD (2) – LUCY

I am dead proud of him. I remember him saying to me, 'It's the most rewarding job I've had.' And that's after all that fast jet flying around the Lake District. And yet he was quite clear, being on XIII Squadron was the most rewarding job he's ever had. I think part of it's working with the personnel, and then the responsibility, but I think part of it's being 'in theatre' and doing what they're supposed to do. Doing good stuff. And killing the baddies.

PROUD (3) – IONA

If I thought he didn't enjoy what he was doing, and he didn't do a good job – and I can see how proud he is of what he does – then I wouldn't keep going.

CHANGED (1) – BRETT

She'd go to work, and she'd fly whatever mission needed doing, because she was in that bubble. Sometimes she'd come home and she would still mentally be in that bubble. She couldn't switch off. She would talk to me about whatever she was doing that day. If people had been killed.

For her, it was good to talk. It was probably easier for her to talk to me because I had worked with the MoD on the intelligence side. Some partners may not be able to go home and talk to their other half, because they don't understand it. But sometimes she could definitely be down for quite a while about things. It did affect her. And when she did stop talking about work, she wasn't able to talk about anything else.

CHANGED (2) – EMILY

It has changed him, but I wouldn't necessarily say for bad, forever. I think it has been good for his confidence because he is good at it. But he's got a shorter fuse now, with the kids and stuff. He hasn't got the patience he used to have. And it's completely understandable. Mind you, I haven't either.

He used to be really chilled out and nothing much would faze him. He rants a lot more now. I suppose his stress threshold at home is lower because the stress levels are high at work. It's difficult to just flick a switch and be cool, calm and collected as soon as you walk through the door.

I wait for him to come home and I'm thinking, *If he goes straight for a beer, then it's been a bad day.* Before he even comes in to say hello, if he's got that beer in his hand I'm not even going to ask how his day was – there's no point. But he will tell me when he is ready, which I think really helps from a relationship point of view. I don't have to ask and I can understand why he's being like that. I think I'd get very frustrated if I just got a moody guy home, with no explanation. I like to know what's happening. I like information.

I think people on the squadron just do need to talk, but you have to have the right kind of personality to do that. You can't just change your personality. My husband is a talker. I'm not, but I'll listen. I think if it was me doing his job I wouldn't be talking about it.

MY HEART ACHES – GEMMA

There is a very recent event that will stay with me forever. On 1 October 2017 there was the mass shooting from the Mandalay Bay Casino and Hotel in Las Vegas. I woke at 3am for no apparent reason and checked my phone for the time. I had a notification from the Sky News app with information on the shooting. I leaned over and checked Fraser's phone to see that he had received calls from the squadron, checking we were all safe. I then spent the next two hours scrolling through Facebook and British and American news sites trying to find out the latest information.

The next two days were emotionally hard; I found myself driving around Vegas crying to radio updates as more information on the victims was released. Vegas felt different, it wasn't angry or hateful to me, just simply sad – and I was grieving with it. I felt compelled to help the victims but having just had an operation I knew I couldn't be involved in lifting the donations that had been received. Instead, we did as we do every day: Fraser went off to 'save the day' while I looked after the kids.

He volunteered at a local charity taking in donations. We wanted to give blood but being British we were not accepted due to the fear that we may still have mad cow disease.

On the night of 2 October, I remember getting into bed and thinking about the victims, their families, their fear at the concert, the panic, confusion and the aftermath. Then a very clear and unexpected thought came to me. I was struggling with emotions created from seeing the loss of innocent lives on this one particular event and yet my husband deals with this every single day. The guys at Creech and Waddington do their job knowing that the people they watch daily are involved in a vicious war that sees the loss of innocent lives at the hands of monsters. Our guys may see those murders for all I know.

It was a light bulb moment that I will remember always.

I again spent several hours thinking this over. I'm sure I need sleeping tablets. The next morning I couldn't keep the thought to myself. I messaged the other squadron spouses:

> There are many emotions that burn within us in times of tragedy, I personally feel completely heartbroken. My heart aches for the families and friends that have lost precious loved ones and those considered 'lucky' who will have to deal with the memories for the rest of their lives. This tragedy will never leave me and I am sure many of you too, and that alone is an overwhelming thought.
>
> But then I think of what XIII Squadron & 39 Squadron do every day. My husband, your spouse,

> our friends; they deal with situations not dissimilar to Vegas on a weekly basis. Knowing and seeing the loss of innocent lives at the hands of cowardly, evil people. I can only imagine the emotion that burns within all of you constantly. The heartbreak, anger, determination and solidarity you must feel and return to without question each night... and for that I thank you with enormous pride!
>
> Call me sentimental for speaking out but I believe we must always make our voices louder than those who seek to do us harm!

I have tears in my eyes reading this again. I hope if nothing else my sharing this with you portrays the emotional investment that I, and I am sure many other spouses, have in our men and women.

The film *An Officer and a Gentleman* – and many others of its ilk, even *Top Gun* and *Independence Day* – have a lot to answer for. The glamorising and romanticising of the lives of military personnel, especially aircrew, offers a Hollywood fantasy of the idyllic lives of those whose partners are in the forces. While high-voltage action sequences light up cinema screens for hours on end, the reality of being close to someone in the military frequently sits in the background.

Every military family develops its own catalogue of stories that will be repeated and repeated, and eventually come to define the family history. Most of the stories involve either the capricious posting gods or things that happen while the military half of the couple is away on duty in some far-off place. When both partners are in the military, the catalogue of stories expands even faster.

In my family, the most commonly reprised incidents are (with the unstated assumption that they are all my fault): electrical equipment, heating or other domestic disaster; car breakdown; the death of the guinea pig when I was away; missing four birthdays in a row (thirteen years ago – still not forgotten or forgiven); the 2am phone call from my wife, when I was thousands of miles away in another country, to tell me there was mouse in the bedroom; and many more. It was part of an almost predictable routine.

Conventional RAF, Navy, Marine or Army families or couples go through a regular pattern of separation, typically for four to six months, every two years. Many others will be away much more. There is a well-established pattern of behaviour.[39] During the build-up to the separation there can be more arguments than usual, or fewer if the couple wants to pretend everything is OK. Couples start to detach from each other emotionally, eventually getting to the point where they both welcome the inevitable departure. After the separation there can be feelings of guilt, anger, loss and vulnerability. I know some people who put photos in drawers to avoid daily reminders of the absence. Then things settle down, briefly, before the homecoming looms on the horizon. Children get especially wound up at this point. An idealised homecoming reunion is anticipated, but in reality is often underwhelming. Both partners will have changed to some degree – especially those who have been in war zones – and need to adapt their expectations before, in a perfect world, things settle down.

But – and it is an important 'but' – for some, the gap between home- and war-life is too big and they find it impossible to mentally adjust. Those who have been away often do not want

---

39 An official example can be found at http://www.royalnavy.mod.uk/welfare/deployment/the-deployment-cycle/the-emotional-cycle-of-deployment, accessed 1 December 2017.

to burden their partners with details of bad, even horrific experiences.

In a Reaper squadron the pattern is different, except when the pilot or SO is sent out to the LRE in some country in the Middle East for a few months. There, they spend their days getting Reapers airborne, then landing them again after they have been flown on missions via satellite from the UK or US. Most commonly, the pattern of life for Reaper families runs in a two-part cycle.

The first part is the twenty-four-hour cycle of daily deployment on operations: anticipation of departure, detachment, separation, homecoming, readjustment and normalisation. The second cycle lasts for nine days – six working days, then three days off (if they materialise). The first day off is commonly spent in an exhausted state, before a brief day or two of calm before the cycle starts again. Having interviewed twenty-four spouses and partners, and spoken informally to quite a few more, the most common observation is that the crew members do not re-adapt on a daily basis, and many do not even adapt during the nine-day cycle. It is an intense existence, which lasts for years at a time, and the real adjustment takes place when it is all over. Until then, the Reaper personnel exist in a form of suspended reality: never fully away, and never fully at home.

POSTSCRIPT

As this book was being written, Lucy and Ed finally got an overseas posting. He did not bring champagne home this time when he broke the good news.

CHAPTER 13
REFLECTIONS

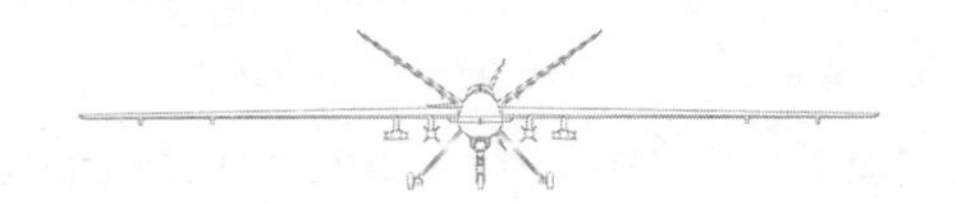

'I HAVE BEEN ON THE REAPER FOR A LONG TIME AND IT HAS CHANGED ME.'

RORY, SENSOR OPERATOR

As this book draws to its close in 2018, I find myself back where it started – at Creech Air Force base, reflecting on life, war and Reaper operations with members of 39 Squadron. These past two years have seen the Reaper Force conduct its highest intensity air operations since 39 Squadron and XIII Squadron were re-formed in 2007 and 2012 respectively.

The ground campaign against IS in Iraq and Syria has descended, increasingly, into urban warfare, and the UN has reported how IS has deliberately kept civilians as human shields in a number of its strongholds. For example, 20,000 civilians were held in Raqqa, as part of an IS scorched-earth campaign.[40]

40 United Nations, 24 August 2017, Note to Correspondents by UN's Deputy Special Envoy for Syria, Ambassador Ramzy Ezzeldin Ramzy, and Jan Egeland, Special Advisor to the UN Special Envoy for Syria, https://www.un.org/sg/en/content/sg/note-correspondents/2017-08-24/note-correspondents-transcript-press-stakeout-uns-deputy, accessed 13 January 2018.

In January 2018 alone, RAF Reaper crews conducted forty strikes, mainly using Hellfire missiles in sometimes extremely confined spaces with little or no margin for error.[41]

Whatever IS atrocities have been reported in the media, the Reaper crews have watched, and much more besides, in real time. On good days they have been able to protect Iraqi or Syrian civilians and the opposition forces as they have advanced building by building and street by street, at great human cost. Sometimes they have had to watch, powerless to intervene, because of the risk to non-combatants. That is a huge number of mental images – some good, mostly bad, some seriously disturbing – to store up for the future.

The experience level of the Reaper operators has evolved markedly. There are a few pilots and SOs who have accumulated 3,000 hours of operational flying, over seven years or more. This equates to many hundreds of missions and dozens of weapon engagements. Fifty or more individual missile or bomb strikes is now not uncommon on the RAF Reaper Force.

Amongst those who are still flying the Reaper, and those veterans whom I have interviewed who no longer do so, what they have reflected upon most frequently with me has been the decisions they have made. But it is not the decisions to shoot and kill some distant Taliban or IS jihadist that they think about most. I have come across hardly anyone who regrets taking any of those shots. Instead, the memories of where they did not, or could not, intervene to help a desperate situation seem to linger the most.

In this final chapter, I share the reflections of a number of the operators, in their own words. Some are more extensive reflections on defining moments, while some are brief, but all say something important. These are, of course, my choices, and I could have

41 The MoD provides regular outline details of such events. See https://www.gov.uk/government/news/update-air-strikes-against-daesh#history, accessed 14 February 2018.

chosen different examples. Beginning with decision-making, the chapter goes on to include insights into the ways in which some people have been changed by what they have seen and done. If these are interesting, I suspect they will not be as interesting as the reflections that will emerge over the coming years.

DECISIONS, DECISIONS

One of the many responsibilities faced by Reaper crews has been deciding when not to fire a missile or not to drop a bomb after authorisations have been given, and the JTAC calls 'cleared hot' in the expectation that they will take the shot. The captain and the crew remain both legally and morally responsible for their actions: they also have to live with themselves and the calls they make, for example when not taking a shot against someone who later goes to kill or harm others.

One of the things that supported a 'no fire' decision from the earliest days of the RAF Reaper Force was a policy direction for zero CIVCAS, no civilian casualties. It became almost an obsession for the Reaper Force and has had a significant impact on how they operate and make decisions.

One of the first 39 Squadron Tactics officers, later Officer Commanding XIII Squadron, explains:

> We knew that zero CIVCAS was unlikely year after year; we were simply determined to make sure that we could always hand-on-heart say that we had used our best effort to achieve the policy direction, right up until the moment that we could no longer influence the outcome. Everything was centred around that policy direction of zero. If the policy direction had been something other than zero, then the military actions would likely have been different.

Against that backdrop, the chapter proceeds with two incidents where the captains and crews decided not to fire when they had been 'cleared hot'. To fully appreciate their actions, remember that push for zero CIVCAS, which, deliberately or otherwise, became an underpinning element of RAF Reaper ethos. In the first example, the crew's judgement retains an element of ambiguity – they do not have their actions confirmed one way or another. In the second example, subsequent events… well, read for yourself.

OFF DRY – JOSH

'Move eyes forward to the compounds of interest,' instructed the JTAC.

'Roger.' As I redirected the aircraft, the SO moved the camera towards the area we were going to watch. There was a nervousness in the request that came over the radio, and rightly so. The coalition unit we were supporting had taken casualties over the preceding few days and were now advancing over open ground to a cluster of buildings in attempt to find the Taliban team responsible.

'Report anything that moves, immediately.'

'Roger.' A first look at the traditional Afghan buildings revealed nothing out of the ordinary. We exchanged the necessary information with the JTAC to allow us to react quickly if something did happen. The soldiers in the friendly unit were potentially in an exposed position if Taliban fighters opened fire, and we were overhead to protect them.

We were primed. A silence came over the GCS as the friendly unit approached the buildings from the west. Any chatter faded away to the minimum comms required –

it helps to avoid any confusion. Suddenly, an adult male emerged from the eastern end of the compound, carrying a weapon. I immediately confirmed this to the JTAC. The final elements of approval came through, swiftly followed by 'cleared hot.' I could now fire.

This was not my first weapon engagement and I would go on to conduct many, many more, but something about the clumsy movement of the target was just not right. The Afghan male was carrying the weapon in an awkward fashion, like a footballer holding a rugby ball, or the playboy uncle that has just been passed the baby. As he started to move away from the area of approaching friendly forces his movement did not have the air of battle-hardened Taliban. Nor did he have the swagger and Jihadi-chic of a recently arrived foreign-fighter, or even the furtive obedience of a $100-dollar militia man-for-hire. And, most importantly, the weapon in his hands was not consistent with the types that had caused such a devastating impact over the preceding few days, or throughout the province.

Part of my brain flashed back to my days growing up in a rural farming community in the 1980s. I recalled a house with a broken back door, and how the farmer used an old, rusty shotgun wedged in position to keep the door closed. I wondered how those farmers, from back home, would react if they found themselves in a war zone such as Helmand. How they would react if soldiers were approaching and they had an unexplained firearm. My deduction was they would react like the guy on my screen: like a panicked farmer in a war zone.

I looked again at the display, the old weapon in the Afghan's hands acting like a wormhole connecting two worlds that would rarely occupy my brain at the same time. I felt sure he was a farmer but I had to be certain in myself, I couldn't let any biases from my childhood cloud the situation. If we got this wrong, either way, the wrong people would die.

Taking an objective 'tick box' view we had an adult male emerge from a compound, armed, as friendly forces approached. The compound was in an area occupied by Taliban that had been engaging friendly forces, successfully, over the preceding few days. It met the criteria needed for a strike, we had all the approvals and authorisation required. But the tiny details weren't right.

I asked my crew, not wanting to lead them to either conclusion. 'What do we think?'

To my relief, they both responded that they felt something wasn't right. The SO was experienced aircrew, with a Tornado GR4 background, and the MIC was a mature army captain who had come through the ranks. I trusted their judgement and it reinforced mine: while this appeared to meet all the criteria, this was not a Taliban fighter.

'Off dry, we assess this is not a valid target,' I radioed the JTAC.

Both in the moment, and on reflection, I entirely understand the expletives that followed over the radio. The coalition soldiers on the ground were the ones in danger and, in their view, we had just increased that risk. We also knew that if the individual on our screen showed any hostility towards them, we would instantly reverse our decision and engage.

Trying to reassure the ground troops was not so easy, especially when you had just withheld a seemingly valid request for a shot. From the perspective of those on the ground waiting for a Taliban fighter to open fire at them was not a good tactic – but this was not a Taliban fighter.

As the Afghan farmer climbed aboard a motorbike parked a short distance away, the radio call came through: 'Can you keep eyes on him until we call in an Apache strike, can you do that?'

So we tracked the farmer along desert tracks and dusty roads, all the time monitoring the radio for the progress of the Apache. We could not drop him from view as he was, after all, an armed individual moving through an area with friendly forces on the ground.

'I'm ten kilometres out from target,' came the call from the Apache… 'seven out'.

The race was on.

'Establishing eyes on target.'

'Withhold fires, approaching collateral concerns,' we interjected. The motorbike and its cargo was entering a highly populated village. The rider dismounted and moved inside a building not knowing how close his weapon had come to being fatal for him.

'Brothers are going to die because of you,' the voice of the JTAC cut us to the core, even in the certainty of our decision.

The comms for the remainder of the mission were professional but uneasy. None of the usual pleasantries were exchanged as we went off-station.

Less than twenty-four hours later, working with the same JTAC and the same crew, we found the Taliban team. There was an anger in the JTAC's voice as he gave us 'cleared hot' once again, as if he expected us to withhold. This time everything was right, and we knew we had to deliver for him. The terrain and conditions made this a far more demanding situation than yesterday's scenario.

The whole crew felt relief as the Hellfire impacted. The dust cleared and we saw confirmation that we had successfully hit the Taliban: the threat was removed. As we stepped out of the GCS, all three of us lifted a shaking hand to each other, as if to compare who had stored up the most tension, waiting for that moment when we could let it all out. The nervousness came not from the technical difficulty of the shot, but rather the dynamic of our relationship with those on the ground. We need to protect them, but do not always agree with them.

I have spent many years searching for the Taliban and then Daesh, using lethal force against them when necessary. It requires a mix of aggression and patience, of calculated professional discipline and empathetic human understanding because every time you hear 'cleared hot' you have to get the next decision right. Failure to do so puts friendly lives in jeopardy or increases the risk to innocents. There are, perhaps, never the 'right' people to die in war but there are definitely the 'wrong' people to die.

OVERSIGHT – MAX

We'd been tasked to track a particular individual and had been following him for several days. Watching and waiting for an opportunity to strike, away from any collateral hazard. There was enough intelligence and legal premise… he was a valid target.

On this particular day, he was travelling on a motorbike from his house to a bazaar. As he left, all of our procedures had been met to the point where we were given clearance to engage him with weapons.

We followed our procedures to look at the surrounding

area, but this particular road was pretty busy. There were all kinds of pressures on the crew because the coalition had been chasing him for a while. He had done some pretty horrendous things to both coalition military and Afghan civilians. He was a high-priority target, and lots of people at different levels of command were watching.

As we went through the day, the Reaper was continually on station. We had already been given clearance to target him four times, and our relief crew had been authorised twice to strike him. So now we're on to clearance number seven and the pressure is building for a strike. He set off from the bazaar towards his house and suddenly the traffic is beginning to thin. Then one of the crew says, 'What's that on the back? Is that *a child*?'

We're zooming in with the camera. We're looking at different aspects. We're discussing it. 'No, it's not a child, it's a parcel. The guy's picked up a parcel from the bazaar.'

'No, it's a child.'

'A parcel.'

'Rewind the tape.' So we rewound the tape to the point where the guy gets onto the bike.

'Yes, it is a parcel. Definitely.'

And that crew was probably one of the most experienced I've been configured with. I was OC Tactics, responsible for developing our procedures and standards, and a pilot instructor. There was also an SO instructor and a MIC instructor. A crew full of instructors within that cockpit and in terms of rank, there was a Squadron Leader, a Warrant Officer and a Flight Sergeant.

I had communicated to the JTAC that we were just double-checking. We'd decided that it was a parcel. But then the Auth in the Operations Room phoned me, since I was the captain. The SMIC had concerns that it was in fact a child.

So we're still having this conversation, and the amount of road we had left before he reached his house was getting shorter and shorter. The amount of time for the missile engagement was rapidly decreasing. It was clearance number seven. The traffic had died down. The JTAC's asking why are we not shooting?

Then I get another message from the Ops Room: 'Look, we're *convinced* that's a kid.'

'Alright, thanks for your input,' I replied.

The established procedure meant that we, captain and crew, made the final decision. Yes, orders could be given to stop us, but the decision was ours.

As a former Harrier fast jet guy, when I was in the Harrier I wouldn't have had other crew members to have a conversation with. I also wouldn't have had people in an Ops Room watching what I was doing and giving me an opinion. If I had been in a Harrier, in those circumstances, I would have pulled the trigger a long time ago. The individual was a valid target and I'd formed a conclusion that that was a parcel on the back of the bike.

The JTAC had another aircraft on station as well, which happened to be British. It was another ISR platform. So he asked those guys to double-check. From the other aircraft came confirmation, 'There's only one individual on that bike.'

So the JTAC said to me, 'Confirmation with another ISR aircraft: one individual on the bike. Cleared hot.' He is expecting a strike, and many people are watching.

But from the Ops Room came, 'No, that's a kid.'

So I was sitting there thinking, *Do I… don't I? I've followed all the procedures. I've double-checked, I've triple-checked and several sources say 'that's a valid target'.*

And what do we teach everybody? 'If there's any doubt, there's no doubt. No shot.'

So I was having a conversation with myself: *Have I had the doubt, treated it and now got no doubt because I've had this confirmatory intelligence from the other aircraft, or is there still a doubt?*

At which point I got on the radio and said, 'We're not engaging.' You can *hear* the reaction on the other end, both from the JTAC and the Ground Commander who's leaning over him, and from everybody else involved. Sometimes you can have these incidents where a lot of the JTACs are Senior NCOs, and quite often they could have a colonel or a brigadier or a general leaning over their shoulder asking, 'Why the hell haven't these guys pulled the trigger yet?'

You could hear it, not by what was said but in what wasn't said. Normally when you make a transmission in military circles, there's quite a sharp response. But when I said, 'We're not engaging' there was a *very* long pause.

Then all I got back on the radio was, 'Roger.'

I could imagine the conversation on the other side.

By now the available road was reducing and there were probably about three or four minutes left before this individual reached his house. Then a nearby British Apache helicopter was brought in. It started its attack run but had to abort for collateral reasons: traffic on the road.

When the guy on the motorbike pulled up at his house, he got off, turned round, lifted up the parcel and placed it on the ground. A two-year-old kid.

Every single person in our cockpit reacted with relief. We had gotten so close to a terrible outcome.

The Auth who phoned me from the Ops Room was a Squadron Leader. But the duty SMIC, who had insisted that it was a child, was an acting sergeant: in effect, a corporal. She had just been given that acting rank and SMIC responsibility, and she had the balls to turn round to me – the former Harrier QWI, Captain of that Reaper, OC Tactics, Squadron Executive Officer (XO) and Second-in-Command of the Squadron at the time – and challenge *my* call. I had the Squadron Warrant Officer in the cockpit as the MIC. So the SMIC has got the balls to tell the XO and the Squadron Warrant Officer MIC that she, as the new SMIC, knew better than them what's on the back of that bike. And *she* got it right. And I'm eternally grateful that she did, and I wrote her up for a commendation.

After the event we went back and reanalysed the tapes and went, 'Where the hell did that conclusion come from, that that's a kid, because that's a package?' Especially given how the guys were interacting around the bazaar.

It was thanks to technology and the oversight from next door in the Ops Room. As a Harrier guy, I would have reached my Law of Armed Conflict decision that that was a valid target and pulled the trigger. The helicopter guys reached that same conclusion. So did the JTAC. Everyone followed the procedures and the processes, doing all the best practice to mitigate risks to civilians according to the law.

We'd done everything within our power: checked, double- and triple-checked, yet in war shit happens. That said, there are far fewer collateral incidents involving the Reaper than manned aircraft. But we had a narrow escape.

FRUSTRATION – MAX

There is a scene in the film *American Sniper* that a lot of Reaper crews identify with. Towards the end of the film, the main character is being interviewed by a psychologist who lists the great number of shots that the lead character has taken and asks him if he is bothered by them. He answers along the lines of, 'It isn't the shots that I've taken that bother me; it's the shots that I didn't take that bother me.'

There was one particular incident where I didn't take a shot. It was in Helmand Province, Afghanistan. When I came into the situation, the aircraft had already been operated by another crew for a whole shift. If the aircraft flies for its full length of day, which could be up to sixteen hours, we'd have a crew at the start, a crew at the end, and we have the 'swap-out' crews in the middle to relieve the duty crews and let them have breaks. I was on the main crew for the second half of the aircraft mission.

I took over the aircraft halfway through its mission, and it was already on station above a friendly FOB occupied by British Army personnel. Within about seven hundred metres of the FOB there was a three-man Taliban team with a heavy machine gun that had previously engaged British forces, and who were hiding now in a Mosque. So we watched it.

The Taliban firing team eventually came out. They had done this three times that day already before I took over flying the aircraft; the off-going crew were furious with frustration. It was a village where the buildings were not very tightly packed. The compound buildings were dispersed. There was a cluster of two or three buildings, then a fifty metre gap; another cluster of two or three buildings, and a further one hundred metre gap. The buildings and compound walls were so arranged that the British forces in the FOB would not have been able to see the Taliban as they ran from cover to cover.

Next to the Mosque there was an area of open ground, about the size of a great big football pitch or more. The door of the Mosque was facing away from the FOB. When these guys came out of the door with their heavy machine gun, they doubled back on themselves around the side of the building. They got to the corner of the Mosque nearest to the FOB and ran diagonally across the open ground. When they reached the far side of the open area, there was a compound wall – an agricultural stone wall about six foot high – with what is colloquially known as a 'murder-hole'. In other words, they'd knocked a couple of bricks out of the wall so they could shove their heavy machine gun through that letter box-style opening and point it at British forces. It looked almost like a fortified sangar position, and they were shooting at British Forces.

As soon as they came out of the Mosque, I was on the radio saying, 'They're out.' The JTAC – and I could hear the frustration in his voice – was being constrained by his Commander at the time. Every time this shooting team emerged, the Commander wanted to see and PID them again. Every time. And each time the collateral damage estimate had to be worked out again for the same location they were firing from. He kept wanting an accurate grid reference indicating where they were. Then he would very methodically make his decision about whether to engage, and then give us clearance to use a Hellfire missile against them. This process was very protracted because, unfortunately, due to some technical gremlin, the FOB forces were unable to see our video feed. If they had been able to see our feed then I think this story would have been resolved by the first crew being given permission to shoot the Taliban team.

The Commander went through this process each time: the Taliban guys come out; we transmit that they're on the move; we repeat the grid of the mosque; we repeat the grid of the open ground. The shooters are now by the firing hole. We're asked to give an estimate of any Pattern of Life or other potential collateral damage, which has already been passed to them, several times. But we've been asked, so we do it all over again; nothing has changed.

By this stage, we could see the gun barrel glowing from all the rounds that were being fired, and we could see the empty cartridges being ejected from the side of the weapon. The gun barrel became so hot that on our screens it looked like a Star Wars lightsabre. In far more professional language than this, we report that the guy's got a 'lightsabre' and that the gun's 'blatting' away; the JTAC obviously knew this and the frustration with the situation was clear in his voice.

When they'd taken their shots, they started running back towards the Mosque. Then the call came through from the JTAC.

'Cleared to engage.'

'Sorry mate, but they're back in the Mosque,' I replied. The aircraft had been on station for a whole shift so far. And while it was the first time I had gone through this process, it was the third time the same series of events had happened that day.

When we got to the sixth time that this happened – my third – I spoke to the JTAC. 'Look, give me a clearance right now. Clear me hot for exactly the same corner of that wall. Then, when they come out next time, I will satisfy all the pillars of the RoE and I will take a shot against these guys – under self-defence RoE for your troops on the ground – before they start shooting at you again from the corner of that field. The permission will only be valid for that location. If the shooters go anywhere else, I will ask for new permission. But there, at that location, I will just tell you I'm doing it. Keep the airspace underneath me clear of helicopters and give me the clearance now.'

'I'm with you, mate. Standby.'

There was a short pause while the JTAC consulted his chain of command. Then the reply came.

'Denied. Please report the next time they emerge and we will go through the clearance process.'

I don't know the reason why the command chain was doing what it was doing. It may be that they'd had a directive to be very restrained in the use of lethal force in that area. Maybe the Commander was new in theatre. Maybe he wasn't used to working with Reaper and needed longer to get to know us before trusting us. Maybe he thought that the ground troops were safe in a very fortified position, that they could hunker down and that the base would absorb the few isolated shots going in. So, for whatever tactical reason, the Commander made the decision that he made.

And that's the point. It's his battle space, and if the Ground Commander doesn't want us to shoot at that moment – for whatever reason – that's fine; it's his call. We will respect the decision. But, God, it frustrated the fuck out of me and the JTAC.

On the seventh occasion we went through the process again, eventually getting the call 'cleared to engage.'

'They're back in the Mosque again,' I informed the JTAC.

'Roger. Standby.' There was a different, hesitant edge in his tone of voice.

Then, 'Man down.'

'Roger,' we acknowledged.

We had watched these Taliban guys firing their rounds all day long, we could have stopped them the first time and on the seventh occasion they hit a British soldier.

'He's been hit but he's alright,' came the feedback from the JTAC, 'we'll get the CASEVAC helicopter in.'

'Roger. Do you want to try for that standing clearance again, so that we can make sure these guys don't interfere with the CASEVAC?'

'Affirm. Standby.' After a brief pause for consultation: 'We are to follow the same procedure. Please inform me when you see any further movement and we will go from there.'

'Roger.' I acknowledged.

Then the firing team came out again. It's the closest I have been in my professional life to pulling a trigger without a clearance. I chose not to. Legally, I would have been fine to invoke self-defence but procedurally, in a piece of battle space that belonged to someone else, who had already given their direction, the professional answer was to follow their direction. That incident is the only time in my life, in my military career, where, after an event, I asked myself the question, 'What would you do differently if you could do it again?' and not been sure of my answer. All these years later I still can't answer myself, hand-on-heart, and say – given that same situation tomorrow – whether I would pull the trigger without clearance.

With the benefit of hindsight, knowing that they shot a British soldier, I would definitely pull the trigger, probably on the second opportunity. Without that hindsight, I think my professionalism would again stop me from pulling the trigger without clearance. I think. Then again, that might just be me resolving the situation using professional logic so that I can sleep at night. That soldier wasn't alright because he got shot. That frustrates the fuck out of me, that scenario. Still, now, several years later, just thinking about it angers me. So when people turn round and say that we're remote individuals who don't have an affinity with the guys on the ground, I think, *utter bullshit*.

HELICOPTER TRAFFIC COP – KEN

If I ever watch a TV programme such as *Traffic Cop* with camera footage from a helicopter, I immediately feel my heart rate start to increase and my mind will start to plan how best to conduct a weapon engagement on whatever person or vehicle is being tracked. Having spent several years looking at the world from, quite literally, a different angle, that perspective is, evidently, ingrained within my psyche and without hesitation is a skill-set that I am subconsciously eager to continue to employ. In all honesty I perversely miss the satisfaction and thrill of thinking on my feet, then planning and conducting a successful engagement; I'm content to say I actually enjoyed it.

YOU CAN'T UN-SEE IT – TOBY

Towards the end of my time on the Reaper, my wife says I was distant, quiet and very aggressive – I will admit that now. I still am to some extent, I still get periods of aggression. I didn't have very much of that during Op

HERRICK when we were operating in Afghanistan, but once IS and Op SHADER opened up in Iraq and Syria, we needed to look after our people a bit more. That's when TRiM came in.

I remember one of the guys had to get psychological support through TRiM. An Iraqi Army camp was overrun by IS fighters and we were above in overwatch. And they were just killing everyone. We saw everything. After an event like that, we'd get all the crew in and have a chat about what happened. We'd always do it the day after. Sometimes we would go for a beer to chat afterwards, just to try and get everyone involved together.

Towards the end of my time, the squadron stopped people from seeing events like that if they didn't absolutely need to see it. So if an administrator, for example, was in or around the Ops Room where the screens show the video feeds, as the SMIC I would close the door and say, 'You don't want to see this.' There are people in those non-war-fighting roles – young staff on the Ops desk who are noting the hours, filling in the Ops sheets, and so on – they don't need to be seeing that shit. Especially when someone's just been cut in half or someone's lost a head. You'll never know how it's going to affect a young person like that.

If *I* didn't need to see it, I wouldn't watch it. But I was the SMIC, so I had to make the fucking video for subsequent review. I had to write up the reports and check everything. So I was constantly getting a mental refresh of everything that happened. The crew or debriefing officer could come in and ask, 'What time did this happen? What time was 'rifle,' when the missile was fired? What time was the strike?' So as SMIC you get to see all of it again and again.

One time I had four strikes in one night, overseeing

two crews in their boxes and watching their video feeds. Sometimes I'd be driving to work at night or in the morning, whenever, and I'd think, *Please God, not today. I'm tired, it's Day 6.* The worst thing ever was having a strike on Day 6 – you'd be tired and you'd have to stay late.

My wife eventually said, 'You need to go and see someone.' Nobody on the squadron was aware of how I was feeling because I was embarrassed and didn't tell anyone.

So I went to see the behavioural psychologist, a really friendly, funny lady. She got me to do some things, like tell her about my experiences. I broke down and cried, telling the story of the attack on the Iraqi Army base. She got me to write letters – not to post, just to help me think through it all. 'Write a letter to your dad, telling him what you think; write a letter to your son, telling him what you think; write a letter to those people who were killed.' She gave me exercises like that, and she gave me some books. I saw her a couple of times but my tour in Vegas was coming to an end.

When I came home to the UK, as soon as we established ourselves here I went to see the doctor and he referred me to the Mental Health Unit at RAF Marham. I was triaged by a psychiatric nurse over five or six visits, where he assesses you, writes about you, learns everything about you and then gives those notes to the doctor.

I asked the doctor, 'Do you think I have PTSD?'

He said, 'I think you have elements of it. But you don't have PTSD because you're not re-living it.' Mine is more of a traumatic memory, I don't have enough ticks in the boxes for it to be full PTSD. I was on tablets – antidepressants – for two or three months. When I was on the tablets I was temporarily medically downgraded, and you can't go near a weapon. I still have the odd dream where I wake up with the 'bang' of a particular explosion.

I'm still seeing the psychologist. There is a way to regress that memory – he says it is in my short-term memory and I am accessing it too easily. He wants to help me with that.

I told my mum two or three months ago. I was with my wife and I just blurted it out. She said, 'I know. I knew when you were out in Vegas.'

She saw that I was more aggressive while driving, very short tempered with the kids, very short tempered with my wife, with her. Just admitting it was the hard part. It gets easier over time but it doesn't get easy.

My wife didn't really want to know about anything I did at work. I was on my own in that respect in Vegas. She wanted to live in her own little bubble, understandably, and not be aware of the things that we saw out there. I saw a lot of strikes. I had six, seven myself that blended into one. You remember your first. Then as the SMIC I was seeing a lot more from the Ops Room.

Although I've been affected, I would still quite happily protect a troop of soldiers on the ground, day-in, day-out. If anyone went near them with a weapon to commit hostile acts I would be happy to intervene.

CHANGED (1) – ERIC

The biggest change was a realisation of my capabilities. When I first started as a MIC I thought, *There's no way a forty-year-old man can do this.* The change in my confidence and abilities became noticeable by the end of my time on the Reaper when I was teaching. It became apparent when a new Squadron Leader pitched up and I could tell that he was in the position I had been in three years earlier. He was absolutely gobsmacked and said, 'I can't do that.'

He was watching me monitor six screens, simultaneously communicating while watching the main screen to give advice in the cockpit. Communicating on the radio. Taking that information and instantly allocating a priority.

I was doing all this and there would be a kill chain on the radio, while MIRC – secure internet – chat was also going on, plus, using my expertise with RoE, the legal side. And then to later play back the video where you are so cool, calm and collected. I was thinking, *that's impressive*. And that was me. Brilliant experiences.

CHANGED (2) – RORY

I have been on the Reaper for a long time and it has changed me. It's difficult to explain. I think that my wife, April, is very supportive. I guess I am more uptight a lot of the time. I find that I have a shorter temper sometimes. I get irritated and I flash at small things that I probably wouldn't have done before. But I think April understands why that is.

Little things that might seem petty – even to me – set me off. Sometimes I just can't stop myself. I can feel it almost welling up inside me: this frustration. Not necessarily anger, just frustration. I don't know where it comes from, and I don't know why. When I first met April, and before I joined the Reaper, I wasn't like this.

I'm never raging and screaming and shouting, never like that. I just get frustrated by things. Now I just walk out of the room a lot of the time. I probably don't say 'sorry' as much as I should. I take myself off and when I come back

> I am generally OK. It's a strange one. Sometimes I think, *Shit, I shouldn't have done that.* I definitely think I'm tired. If anybody on the Reaper fleet says it doesn't affect them, then they're lying. It does. It has to.
>
> But I really enjoy what I do and I don't think I would change anything about the last five years with the Reaper. Five years is more than enough in one go. I am well aware that I am probably six to twelve months overdue a rest, just physically and psychologically. That said, I'm excited about what's to come in the future with the Protector – the successor to the Reaper. And for non-commissioned aircrew, I don't think there's a better job in the Air Force at the moment.

These reflections, and even this book, are not the final word on the RAF Reaper Force and the people who belong to it. It is a foreword, a brief glimpse into the human dimension of remote warfare which, at best, provides some initial insights that pose as many questions as they answer. The quoted reflections on different aspects of Reaper operations, together with the previous chapters, highlight how profoundly immersive the human dimension really is. Reaper crew members have taken life and death decisions on a weekly or daily basis over periods of years. Decisions to shoot, or not to shoot; decisions to keep watching, then shoot at another time. At times of high-intensity operations, as in the fight against IS in Syria and Iraq in late 2017 and into 2018, those decisions have been made on an hourly basis. Then there is the watching: observing horrific events that cannot be unseen. Executions, beheadings, IEDs, attacks on friendly forces and much more.

Everyone is affected to some degree, although it varies from person to person; it would be more concerning if anyone was unaffected. Some Reaper operators cope reasonably well throughout their time flying the aircraft, and conducting both lethal weapon strikes and seemingly endless hours of surveillance of far-off enemies. A few individuals have conducted up to sixty and more strikes against Taliban or IS fighters, resulting in death tolls that run into three figures.

As I complete the final elements of research for this book in February 2018 at Creech Air Force Base with 39 Squadron, the conversations regularly turn to the future. The questions that emerge are significant, both for the RAF – and any other air force that uses RPA like the Reaper in war or counter-insurgency – and for the people involved. At a human level, is there a limit to how long high-intensity operations, like the urban warfare against IS, and the killing of enemy fighters can be sustained? As piloting and weapon proficiency improves to the point where the margins for error are reduced to a few feet in some instances, is there a risk of over-confidence or complacency? Does prolonged exposure to killing desensitize those who watch and carry it out? Should ethical questions like 'Is it right?' matter as much as 'Is it legal?' when it comes to lethal strikes?

After almost hundred interviews, and hundreds more conversations over several years, my answer to all of these questions is a qualified, 'Yes, probably.' However, nothing is that simple. There are a very small number of pilots and SOs, probably fewer than ten, who have performed at the level of elite athletes (think Olympic shooters) for seven or more years. Meanwhile, many are physically and emotionally tired after two or three years, and, if they stay on the Reaper Force, need to be thinking about doing something else after five or six years.

The Reaper Force is still too new to be able to fully understand all of the factors that come into play and how they interact. But there are some obvious considerations. Operational tempo, the sheer number of hours worked and the shift patterns involved provide the starting point. Then there is the nature of the operations. Reconnaissance and surveillance – watching – for days and weeks on end can be mind-numbingly boring and low stress. In contrast, frequent weapon firing under intense pressure can be repeatedly adrenaline-inducing. And, for some, utterly draining afterwards.

Which brings in the second factor: individual physical, mental and emotional resilience. At one extreme are those who thrive on the challenges and can't get enough. They almost have to be levered out of the GCS for a break, or at the end of a shift, if there is the chance of striking some key IS or other target. Meanwhile, a very small number struggle with pulling triggers or guiding weapons onto targets. In between, the majority just want to do a professional job to the best of their abilities.

I would offer my own lifelong commitment to lifting weights as an admittedly limited comparison. For many years, I was a power-lifter who could squat and dead-lift almost three times my bodyweight. The idea of missing a training session horrified me; even in my fifties it still does, even after the strength has waned. The endorphin release and the adrenaline kick have been my drugs of choice for decades. I find it impossible to understand how anyone cannot love the agony and repetitiveness of lifting weights. But apparently there are people in the world who don't even enjoy going to the gym. Why do I? Hard-wiring. I have previously alluded to my squeamishness. Again, hard-wiring.

Military history has many examples of individuals who could not pull a trigger to kill an enemy, while others have been willing, some even keen, to do so. I have encountered those differences in Reaper operators. Responses to extremes of human behaviour – endurance sports, strength sports, killing in war – are individualised, and training can only go so far in altering them. You can't defeat hard-wiring.

Then there is the moral dimension. Several operators have spoken about how IS is an 'easy' enemy to fight because of the barbarity – beheadings and other killings – they have witnessed through the live video feeds. That strong sense of moral purpose heavily motivates. Conversely, a lack of moral purpose can be demotivating. For RAF Reaper personnel, the pursuit of a policy of zero CIVCAS has been the moral foundation of what they have done for a decade. Chapter 5 looked at where it went horribly wrong. Any change to that policy would likely cause problems for a significant number of crew members.

I conclude with one final challenge that all Reaper personnel will face: leaving it all behind when their time is over. The intensity of the work is so great, and the stakes so high, that many find it difficult to readjust to 'normal' life, either within the armed forces or as civilians. Nothing they ever do in the future will be as demanding, stimulating, gut-wrenching or fulfilling as those moments when a missile is in the air and the difference between success and disaster, life and death, is measured in feet and inches.

'I still think about the Reaper every day,' Zach said in a whisper, as though he was making some kind of confession. For a few moments he was lost in his own thoughts.

By his own admission he was exhausted by the time he flew his final mission. He also admits – and his wife reinforces the point – that flying the Reaper has affected him. He fired more than thirty missiles and bombs in Afghanistan, Syria and Iraq, and has a whole mental catalogue of images as a result.

'But I wouldn't take a single one of those shots back,' he tells me, adding, 'I would do it all again.'

I ask him why, when he has clearly paid a price.

'Because I made a difference.'

EPILOGUE

REMEMBERING

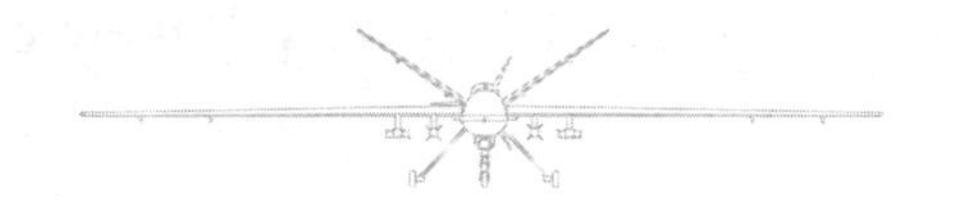

'DO YOU WANT ME CHECK IT?'

CORPORAL MATTHEW T. RICHARD, US MARINE CORPS

The greatest responsibility of a US Marine is not to support and defend the Constitution of the United States. That is his *first* responsibility. Neither is the greatest responsibility to deploy to a distant battlefield, and protect and encourage fellow Marines while overcoming their common enemy. That is his second responsibility. No, the *greatest* responsibility of a US Marine comes after the battle is fought and the sacrifices made, when a flag-draped coffin bears his brother-at-arms on his final journey in this world, and he must somehow find words to say about the man he fought beside and find the strength to speak them. And then to remember.

On this day, 1 July 2011, that responsibility would fall to squad leader Marine Sergeant Seth Hickman, at Patrol Base Lambert in Helmand Province, Afghanistan. Earlier, he had gathered his squad together so that they could all contribute words to the eulogy that none of them wanted to write and none of them wanted to hear. In the heat of the desert air, Sgt. Hickman stepped forward to pay tribute to their fallen fellow Marine, Corporal Matthew T. Richard.

> Some of you knew him as Cpl Richard but we knew him as 'Dick'. He didn't take to the name at first but eventually learned to deal with it.
>
> There's a lot that can be said about Dick; he was a good guy, a better Marine and the best at being a friend. Whether it was his confidence, sense of humor or plain old belligerency, he drew you to him and it didn't take long to see that he was one of us.
>
> A kid at heart but always a professional, when it called for it. You could always count on him to watch your back. He would jump at the chance to put himself between his brothers and danger, time and time again. He was a true warrior.
>
> Slow to anger but quick to complain, serious one moment, all smiles the next. It didn't take long to realize that even his rare moments of anger had a hint of humor. Dick loved his work, but the better you knew him the more you would realize it was because it always gave him something to bitch about.
>
> Outspoken as hell, he would take any opportunity to let you know you were screwed up. Yet easy to talk to, he was always willing to dole out advice even if he had no clue what he was talking about. One thing is for sure, he wasn't afraid to let you know he was right and you were stupid. But that's why we loved him, no hesitation. The cornerstone of his personality, no matter the situation, you could always depend on him to put himself out there and do his best to resolve it.

To Richard there was nothing more important than family. We can only hope that in the end he saw us as we saw him: Our Marine, Our friend, Our Brother.

May you rest in peace Brother!

As he finished, red eyes were wiped by men who would rather face an enemy's bullets than the strength of their grief. They would carry this moment with them forever. In time, the surviving members of Matthew's squad would create their own permanent memorials on their bodies. They would get a tattoo each, showing him on patrol among the poppies of Afghanistan: erect, alert and peering into the distance, cigarette in his mouth and his Camelbak pack slightly dishevelled.

The tattoo also shows his dog tags. Military personnel carry two dog tags. One will be removed from the body after death in order to notify the chain of command, while the other will remain with the body for identification.

Back in Iota, Louisiana, Matthew Richard's parents, Jeff and Alicia, and siblings, Joshua and Laura, were trying to come to terms with a loss that no parent, no sibling can ever come to terms with. Presidents, prime ministers and other politicians often talk of duty and sacrifice in war. But those words – 'sacrifice', 'loss' – are too small to encompass the depth of human devastation felt in every affected family, every community.

The physical absence at family gatherings is reinforced by a sense of the person who *should* be there, but is not. Family rituals are broken and never fully recover. School reunions will always be incomplete. And a squad of Marines will go back out on patrol in the dirt and heat carrying unspoken memories and feelings alongside their weapons and kit.

Unknown to Matthew's squad in Afghanistan, or to his family or friends back home, on that day he was being remembered by a British airman who had never met him. An airman who had seen Matthew only once, and who still thinks of him almost every day.

9 JUNE 2011 – 39 REAPER SQUADRON, CREECH AIR FORCE BASE, NEVADA

There is no way to quantify a 'normal' day when working as crew on Reaper, but 9 June 2011 started as normal as can be. Ross was the MIC in a 39 Squadron crew based at Creech Air Force Base. Their task for the day had several strands. The first was to provide reactive surveillance scans of an area of Helmand Province to provide intelligence for the American Marine unit they were supporting. They were also to be on hand to support the field-deployed Marine patrols should they get into a TIC (Troops In Contact) situation.

A TIC could take the form of an ambush on a foot patrol or a vehicle-mounted patrol. Or, there might be a direct assault by the Taliban on any of the Marine fixed-position bases that were dotted around. Even with the close air support provided by British and American Reapers and other air-to-ground attack aircraft and helicopters, the Taliban still had a number of successes. They mainly employed classic insurgency tactics: strike as fast and as hard as possible; inflict what damage they could; then withdraw before the US-led coalition could deploy its overwhelming force. And the US Marines could deploy overwhelming force quickly and aggressively.

To protect against an attack, Reaper could provide eye-in-the-sky overwatch for the Marines on the ground so that they knew where the threats were coming from and how to react. It could also mean direct action using missiles or bombs to protect the Marines, or a combination of the two.

The first part of the day's sortie involved scanning an area and looking for overt Taliban activity, such as carrying out armed patrols or laying IEDs. Ross would check and re-check images and video footage if there was the slightest suggestion of anything interesting or threatening. But so far, nothing.

Everything seemed pretty quiet when Ross's crew received a radio call to 'move eyes' to a location where a US Marine vehicle patrol had struck an IED. They were to provide overwatch. There were no casualties reported, testament to the effectiveness of the mine-resistant vehicle, but whilst they awaited recovery of the damaged vehicle, the Marines had, in the main, left their vehicles and had taken up defensive positions off to the side. No one wants to sit in a stationary vehicle, in the open, on a raised road. IEDs like this were sometimes a precursor to a sniper taking a few shots at static targets or they could be followed by an ambush or a larger attack. Just in case, the top-cover gun positions of the vehicles were manned.

Once Ross's crew had eyes on the Marines, they began to ascertain the make-up and layout of the now-static formation and where its perimeter was. This assessment would enable them to work out the areas from where threats were more likely to emerge. At the same time, they would be assessing the Marines' posture: are they hunkered down, indicating they feel threatened or that they are reacting to incoming fire? Alternatively, are they relaxed? These visual indicators are the catalyst for what type of cover or search might be required and the urgency with which it is carried out. Such visual clues are especially important in situations like this if direct radio contact isn't possible because of a lack of frequency sharing or suitable equipment on the ground. This group appeared fairly relaxed. Not bad considering the possible alternatives.

With no immediate threats apparent, Ross, his pilot and SO

continued to scan the area, expanding their search from the initial grid coordinates. Not too far away, they observed a Marine foot patrol near a walled compound, some standing, some sitting. There did not seem to be any urgency, any obvious threat against them. One of the Marines walked from the group and along the compound wall. His movement caught Ross's attention, and for no reason he could explain later, he asked the SO to follow the individual who had peeled away from his colleagues.

From altitude, there was no way of knowing if the Marine had got bored and wandered off, whether he was 'taking point' – leading the way – on the patrol and simply looking ahead, or if he had been ordered to check something out. The Marine turned the corner of the compound and edged along the wall before stopping just short of some kind of doorway or opening. Curious as to what he was doing, Ross asked the sensor operator to zoom the camera in on him. They were watching a squad from Golf Company, based at Patrol Base Lambert, Helmand Province.

9 JUNE 2011 – PATROL BASE LAMBERT, HELMAND PROVINCE, AFGHANISTAN

For several days, whenever the Golf Company squad headed out west of Patrol Base Lambert, they took incoming sporadic fire. It would give the shooter too much credit to call him a sniper – the shots were never close enough to hear the crack of the round nearby. But it only takes one round…

The shots were coming from the south. Sergeant Seth Hickman, the squad leader, and his squad had the likely source narrowed down to a small area: a couple of houses and compounds almost one mile away, adjacent to a road that ran south from Lambert, with a canal running parallel.

Hickman came up with a plan to deal with the threat. He would split his squad in two: four Marines and a corpsman (they provide medical support during operations and carry specialist field medical equipment) would take up an overwatch position. Hickman would lead an assault team of nine Marines, two Afghan National Army soldiers and an interpreter south down the canal, remaining as unobtrusive as possible. Meanwhile, the overwatch team would make their way as stealthily as possible to a point several hundred yards west of Lambert and wait. When the assault team was in place, close to the compounds, on Hickman's signal the overwatch team would start walking overtly towards the base to draw fire. When the shooter or shooters started firing at the overwatch team, the assault team would be in position to identify them and return fire.

Hickman and his team had moved as covertly as possible a few hundred yards down the canal. Up front, LCpl Rosperich and his metal detector were sweeping a clear path for them to follow. Behind, every man scanned his own visual arc, ensuring nothing and nobody was missed. The green foliage surrounding the canal provided some cover. On their left was a berm – a man-made embankment – about 8-10ft high, while on the right was a road at the same elevation. From the air it resembled a green snake against the rocky, sandy terrain. On the signal, the members of the overwatch team started moving from their position towards the base. Everyone was more on edge than usual. Hyper-alert. More focused. Incoming rounds could start at any moment.

They registered a kid playing on the road running parallel with the canal. Was the boy aware of them? The kid stopped what he was doing and headed towards the nearby house and compound on the west side of the road.

As they walked further and further along the canal it got

much quieter. Rosperich soon got a hard hit on his metal detector. Richard was beside him, and the engineer, Cory, ran up quickly. He dug up what turned out to be some metallic trash that had been buried for who knows how long. They started heading further along the canal, stopping for a moment in a small depression to take a breather. Richard wanted to get a look ahead so he and Rosperich ran 25 yards or so further along to see if they could get a glimpse of anyone hanging around in the surrounding fields.

A few minutes after the overwatch team started to move everything was quiet. The assault team members were scanning from middle-to-far distance. Then, a deep, massive *thmmmpp*. The stillness was shattered by a huge explosion right next to them by the canal. Dust and foliage filled the air.

Up ahead, Richard and Rosperich were sitting alone and had been watching the fields to the east for about five minutes or so when they heard the explosion. Richard jumped up and gathered his rifle. Rosperich grabbed his arm.

'Hey man, I don't hear any yelling, we should keep an eye out for a trigger-man,' Rosperich reasoned with him.

'I have to go back,' Richard replied, shaking his head and shaking Rosperich's hand free of his arm. Rosperich knew that if there was a chance that one of Richard's Marines was hurt there was nothing he could say that could keep him away. Richard always had *his* back, so Rosperich grabbed his rifle and ran after him.

Hickman and his team felt the blast as much as heard it. LCpl Hunt was the nearest member of the squad, close enough to be knocked over and briefly lose consciousness. Momentarily, everything seemed to pause as his brain tried to process what had just happened.

'I'm OK,' Hunt called to Hickman as he came round.

Hunt felt something hit his Kevlar body armour. It was a kick from a dusty, standard-issue desert boot. He looked up to see Richard, who had a relieved grin on his face.

'Just checking,' he said.

Hunt could see Marine Hollar, the radio operator, squirming around on the dirt a few feet away, holding his right hand. Hickman, Hunt and Hollar had been about 10ft or so from the seat of the blast. Hollar took a sharp sting in the hand. Shrapnel. Blood quickly started oozing. Everyone was snapping into action, taking whatever cover was available and adopting a defensive posture while scanning for further threats. Corporal Gordon attended to Hollar and patched up his hand. Hunt went over, took a knee – knelt down – and asked Hollar how he was. The wound was not as bad as initially feared. Hollar's hand hurt but still worked. He would keep going.

Had the pot shots been a ruse to entice a Marine patrol to head down the canal? From the nature of the explosion they assessed that it had been a remote control directional fragmentation charge improvised explosive device (RCDFCIED). The good news was that it had not been triggered by one of the squad standing on a pressure plate. It had been set off remotely. Whoever detonated it had been slow in reacting and missed the chance to detonate the IED at an optimal time to cause maximum carnage. Instead, it only managed to scare a bunch of the guys and give Hollar a small piece of shrapnel in his hand. The bad news was that someone had anticipated their presence. Rosperich noted that when Hollar held up his bandaged hand to show him, he seemed strangely cheerful.

The squad leader automatically went into post-blast assessment mode. Secure the area. Who else was hurt? Are there any more threats? Send a report for an Explosive Ordinance Disposal (EOD) team to come in. They would want to know all of the details about the blast. Did the IED have the signature of any specific bomb maker? Was this a new development and how could it affect other patrols?

The two nearby houses and their compounds needed to be secured: one on the east side of the road ahead and the other on the west. Hickman sent Cpl Gordon to lead a team comprising LCpl Prenkert, LCpl Corey, the two Afghan National Army soldiers and the interpreter to clear the house on the west side. Richard's team, including Rosperich, who was on point with the sweeper and checking for land mines or IEDs, Hunt and LCpl Thayer, were sent the very short distance to the house to their left on the east side of the road. Hickman was happy to let Richard take his team off and do what needed to be done.

Richard did not tell Hickman every little thing he was doing, and that was OK. He had earned that trust over a long period. Hickman had been Richard's Infantry Training Battalion Instructor at the School of Infantry. He had spotted the potential of the young Marine early on; potential that was sometimes obscured by a lack of enthusiasm for doing exactly what he was told. Hickman had seen enough good stuff to fight to get Richard assigned to his squad. He had also seen enough to give Richard the unflattering nickname, 'Dick'. *You can train someone to follow orders – he'll come around*, was Hickman's thinking. But character cannot be trained and Richard had plenty of it. Extrovert. Gym rat. Knows everything about nothing and nothing about everything – and always happy to share it. Nobody really knows why it is that some people will take a step forward when confronted by danger, while others want to stand still or step backwards. Richard liked to take things on and Hickman respected him for it.

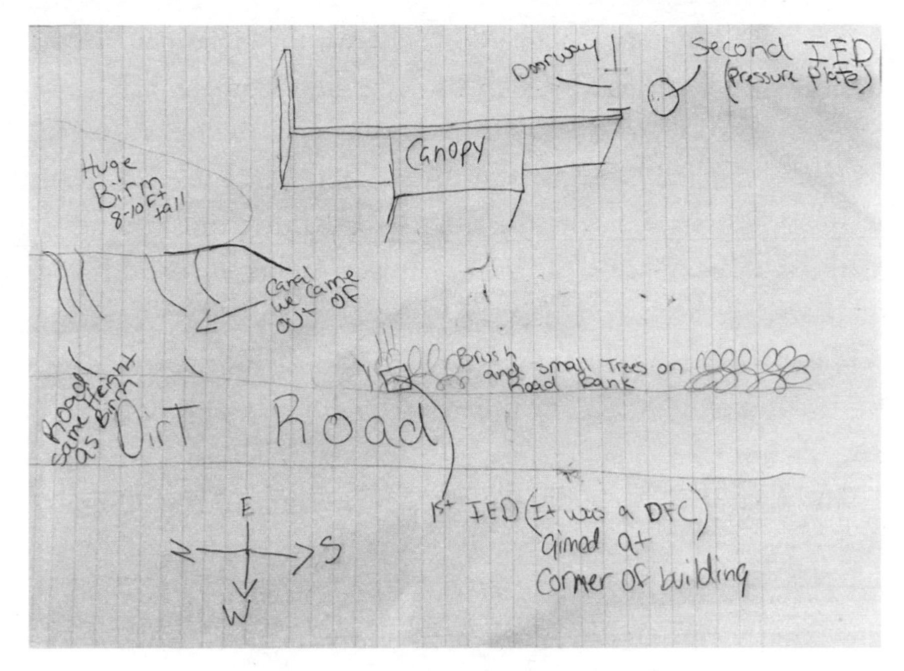

Sketch of IED blasts – Hayden Hunt

Richard's team edged down one side of the compound wall. Rosperich led the way, clearing the path with each sweep of his mine detector, with Richard just behind him and Hunt and Thayer following on.

Hollar was nearby with the radio. Hickman would have been around 25ft away, scanning round, keeping an eye on Gordon's team as they headed to the house and compound on the west side of the road. Rosperich got 6ft past the corner when he got a 'hard hit' – a solid metallic beep from his detector. It told him there was something suspicious beneath the surface. He could also see that the ground had been disturbed. He felt his stomach clench: it looked wrong and made him feel edgy. He didn't want to go poking around this one with the Ka-Bar – standard Marine-issue combat knife – Richard had lent him. He turned back to Richard, who was wearing his goofy grin. He could sense Rosperich's unease.

'Do you want me check it?' Richard asked, with his usual upbeat demeanor.

'Yeah man, I don't want to. I have a bad feeling about it,' Rosperich replied handing him the Ka-Bar.

Rosperich took Richard's rifle and rounded the corner to lean it against the side of the building with his. Rosperich and Thayer were both out of sight of Richard around the corner. A few feet away, Hunt was in the open, able to see along both compound walls, in line-of-sight of Richard, Rosperich and Thayer.

Hunt watched Richard take his Ka-Bar knife and start edging round towards the point where Rosperich had got the hit on his detector. The idea was that when he got close, he would get down in the dirt and inch forward, gently and repeatedly easing the Ka-Bar into the soil ahead of him, to locate whatever had given Rosperich the hard hit.

The Marine who had peeled off from his colleagues now filled the screens in front of Ross, his pilot and the SO. The Marine's posture and slow movement suggested he did not feel under immediate threat. The three members of the Reaper crew were looking at him from above and slightly behind. As the Marine looked down towards the ground in front of him, Ross stared intently over his shoulder, trying to work out what he was seeing. Then the Marine bent down…

And silently the screen exploded.

The Marine at the seat of the explosion was lifted up and back in the direction of the corner he had just walked round, falling where his colleagues were. Although he was alone when the explosion started, he was with his fellow Marines when the dust cleared. The SO immediately zoomed out the camera view. The crew could see other Marines jump into action.

The explosion shook the ground, the pressure wave hit everything and everyone nearby, and the world turned brown.

Hickman was still about 25ft away, close to the road, where he had set up with Hollar. In the fractions of a second it took his brain to register what was happening and for his eyes to focus, he could see Richard falling from what looked like 10ft in the air. He was reappearing into view by the corner of the compound wall that he had walked round.

Hunt was still in the open and wearing electronic counter measures to disrupt radio-triggered IEDs. He registered Richard being blown upwards. It was the second IED explosion to go off near him in a matter of minutes. There was no discernible gap between the force of the explosion hitting Hunt and Richard's left shoulder hitting Hunt's head as he fell back to earth. Whether it was the blast or Richard landing on him, or a combination of both, Hunt momentarily lost consciousness again.

'MATT! MATT!' screamed Rosperich, running to where he had seen him last.

'He's over here.' A faint voice called out behind Rosperich. It was Hunt, prone on the ground.

Rosperich ran to the body but could not see any sign of life. He grabbed Richard's medical kit and got to work. All those hours of field first-aid training kicked in.

When Hunt started to properly come round after a few more seconds, Rosperich was already working away at the right side of Richard's body. Hunt was alert enough to start applying a tourniquet to his left arm. He was also alert enough to see that there was no sign of life – no response.

'Keep going!' Rosperich shouted at Hunt. Hunt would later recall that Rosperich sensed he was either giving up or not fully focused after a concussion. Hunt responded but Richard did not. Rosperich wouldn't stop… couldn't stop trying.

Hickman shouted at Hollar to call it in on the radio. As long ago as the Korean and Vietnam wars, the concept of the 'Golden Hour' developed within field medicine. The better the trauma treatment in that first hour the better the chances of survival. By 2011, that had been refined to the 'Platinum 10 minutes'. The medical evacuation helicopter would be there within minutes.

Prenkert arrived to help and Hickman checked for any vital signs. None. Nobody was yet willing to accept that Corporal Matthew T. Richard, US Marine Corps, had died instantly and could not be brought back, though the thought had started to take hold. They kept going with their first aid.

In the Reaper GCS there was no time to dwell. They were looking to see if there was a trigger-man. There were no direct comms to the patrol on the ground, so the pilot spoke to the Supported Unit while the SO and Ross instinctively searched out wide to look for anyone running away. Or an ambush. Or somebody doing something, anything. There was nothing.

I have just seen an American Marine lose his life. Ross could not keep the thought at bay, even with everything that was going on. Even though he had no sure way of knowing. The death was not confirmed as yet but it didn't stop the thoughts.

It was probably a victim-operated improvised explosive device (VOIED); a simple and dirty, unpleasant weapon that does not discriminate between military targets, civilians or children. They were often laid at or near entrances or paths that a patrol would use.

For Ross, the reality of what had happened kicked in quickly. Shock can take time, and came later, but anger was quick to surface. Ross had once been in the position of those surviving Marines: trying to save an unsaveable life in the Afghan heat, dirt and dust.

Using his available men who were not working on Richard, Hickman started to set up security for the L-Z (landing zone) for the helicopter. Some sophisticated Taliban ambushes would start with an IED or a suicide bomb, but the real target would be the team – especially a medical team – responding by air or road. They had to make sure the helicopter would not be coming into a secondary attack.

After the first blast, the overwatch team had left their previous position about 200 yards west of Patrol Base Lambert, and started moving as quickly as they could towards the rest of the squad. Progress was slowed by their own need to avoid traps. LCpl Carman was using the 'hook' – a lightweight hook on an extended pole that would extend several feet in front of them – to check for trip wires. Meanwhile Cpl Henderson had the team's 'sweeper' and was checking for mines or IEDs under the surface.

They arrived shortly after the second blast to find several of the other team working on a prone Richard. Tang – the Corpsman with additional medic training – got to work as well, while Carman and LCpl Lentz took up protective positions. When Henderson first caught a glimpse of Richard on a stretcher, he allowed himself to imagine that he would soon be holed up in a hospital, getting patched up, and preparing for the trip home. He had seen it happen to someone else. When Lentz broke the news to him, reality started to sink in and his comforting daydream evaporated.

By the time the helicopter arrived, Richard had been unresponsive for more than ten minutes. He had died instantly, and in the moments and minutes afterwards he had been surrounded by his friends, his brothers-in-arms. They had done all they could but were not able to bring him back.

Hickman and Tang helped to carry his stretcher to the

helicopter. Hickman had never lost a man under his command before. He would never lose another.

Once the EOD team arrived, with at least one human intelligence specialist, the compounds on both sides of the road were searched and the occupants questioned. IED-making equipment was found and two suspected bomb-makers were detained. They would not be laying IEDs for anyone else any time soon but the cost had been huge.

When the squad eventually arrived back at Patrol Base Lambert, the Company Commander and 1st Sergeant took them aside and briefly spoke to them. Nobody could take in what was being said. 'Keep fighting hard. Keep focused,' was the key message. Mentally, they were all fading out to varying degrees. After walking back from the IED site, energy levels were dropping, the extreme and prolonged adrenaline surge was wearing off and shock was starting to set in for most of them.

A world away, Ross was driving home towards Vegas after his shift ended. He began to replay the death over in his mind. *What did I miss? What could I have done to have stopped it?* The answer was nothing. He had only been able to watch. When a message came through later from the Marine unit thanking them for being there, he felt hollow.

What made it worse was the time Ross had spent previously as a Casualty Notification Officer (CNO). He thought of that process beginning just now and the CNO who would have the unenviable task of telling the family that their son, brother, father, husband had died.

Nobody who has ever had to give that news to loved ones ever forgets the emotion involved. The importance of how that moment is handled can be overwhelming. Get it right – share the terrible news with dignity and respect – and the family will remember it forever. Get it wrong – incorrect name or incident details, disrespectful or undignified tone – and the family will never forget. More than a few battle-hardened officers and soldiers have stood with a chaplain at the gate to a next-of-kin's house, frozen to the spot and unable to take the steps that will end with a family entering a living nightmare than will never truly end.

Then it dawned on Ross: *I knew of his death before his family did. They are currently oblivious, and will be woken with the news.* The thought pained him. Physically. He arrived home drained, drove into the garage, closed the door and, for a short while, cried.

9 JUNE 2011 – IOTA, LOUISIANA, HOME OF JEFF AND ALICIA RICHARD, MATT'S PARENTS 5.45AM

Alicia heard the doorbell ring and the dog barking. She elbowed Jeff and told him someone was at the door.

'It's just the dog,' he replied.

Jeff rolled out of bed and started for the kitchen when Alicia heard the doorbell ring again. During those split seconds between Jeff getting out of bed and the doorbell ringing, her mother's intuition kicked in and her thoughts went directly to Matt. But then she dismissed them, reasoning that someone must be stranded and in need of help. It only took seconds to realise that her first intuition was right. Several times during his deployment to Afghanistan she had a strong sense that he was not coming home.

Alicia heard Jeff yell, and when she ran to him, he was on his knees. He had not even made it to the door. A glimpse of the uniformed silhouettes outside told him what had happened.

Alicia knew; they both knew. They opened the door to four men. A few moments later, Matt's sister Laura came out of her room, she let out a scream. She sensed what had happened. The rest was just a formality.

A tall, uniformed Marine spoke, asking Alicia to confirm that she was Alicia Richard, mother to Matthew Thomas Richard. Her first response was 'No', as if that would make him stop. She has no idea how many times the Marine had to repeat the question before she replied, 'Yes, I am Matthew's mother.'

Alicia still cannot recall what he said, only Jeff pleading: 'When can I get my boy back?'

They listened to the words of their Casualty Assistance Officer, Staff Sergeant Robert Unterburger, but didn't take anything in. Something about going to Dover; about giving them time to let the family know; about coming back in a few hours to talk; about being there for them throughout the entire process. Alicia could see how nervous the Unterburger was and asked if it was his first time.

'It's ours too, and we will get through it together,' she responded.

They felt like they were on autopilot when they called Matt's brother, Josh, in St Paul. *Thank you God, that he was on a Skype call with an old family friend and wasn't alone*, thought Alicia. Then they made their rounds to family and friends to tell them in person. Each time they said, 'Matt was killed in action, IED,' it became harder and harder.

The next day, as Ross drove into work at Creech Air Force Base, he saw the American flag at half-mast. It was a common occurrence that signalled the death, in combat, of a member of the American armed forces. This one was more poignant than usual – he had seen it. A thought crossed his mind: *the most intimate of life's events and I didn't even know his name.* He needed to know about this guy.

On his next day off, Ross searched the online American Service Casualty list for the previous week. There was one person from the unit they were supporting, on the correct date and in the right operating area, Cpl Matthew Richard, USMC, 21, of Iota, Louisiana. *Bloody hell, he was young - I have two sons, not much younger.* Searching further, he found a memorial page to Matthew asking for comments. He saw his face for the first time: short, cropped hair, strong jaw and blue eyes with just a hint of mischief. The Marine somehow became real in a way that he wasn't before. As he read the many messages of condolence, Ross was comforted by the positivity that surrounded the grief, profound sadness tinged with pride.

The next week was a total blur for the Richard family. The following day they left with the CAO for Dover Air Force Base, over 1,000 miles away in Delaware. They met with the chaplain, and waited and waited. Finally, Friday midnight, they were taken to the airstrip where Matthew arrived with two other fallen servicemen. Reality began to set in as they stood and watched the dignified transfer of the coffins. The next day they met more people than they could recall. The one who stood out was the representative from Mortuary Affairs who described what was going to happen, then advised the family that they should not view Matt's body. *How will our family ever get closure?* Alicia wondered.

At Dover, the family had a couple of visitors, a Marine who had served with Matt and who was injured in February, and the mother of another Marine who had not been far from Matt during the explosion. The family stayed in Dover for an extra day so they could make a trip to Bethesda to see one of Matt's brother Marines who had been severely injured by an IED in March. They were greeted there and comforted by several Marine mothers, part of their Marine family. The trip back to Iota was quiet and long.

Over the next couple of days they met and made plans with their priest and with Robert, their CAO. It was decided that they would meet Matthew's coffin at the airport at Eunice – a small, neighbouring town where he had been born. They welcomed him back on a dreary Wednesday morning. The caravan to the airport and back is now all a blur. A family on autopilot. Despite trying to keep things as private and quiet as possible, the streets were lined with flags and the whole community seemed to have turned out. The children stood out most: Alicia's former 4th Grade class and the Boy Scouts in uniform. The family was profoundly moved by the outpouring of love and condolence.

By the following day, Thursday, 16 June, an entire week had passed. The family was still numb, but now fully exhausted as well. The funeral was held in their local church, Matthew's church. The church was full, the music was beautiful, and the priest captured the essence of Matthew through his homily.

The cemetery was packed as they laid Matthew in the ground. His mother Alicia recalls: 'What I remember most is the heat and the distant voices around me, the 21-gun salute and the playing of Taps. I will never forget those. I cannot see or hear either without having a knot form in my throat.

'We realized that although Matt was our child, the entire community had a vested interest in him and his sacrifice. He became our hometown hero, loved and respected by all. But eventually the visitors stopped coming, people went back to their normal daily lives and we were left hanging on to something we just couldn't comprehend.

When Sgt Hickman and his squad arrived back for their own

homecoming, Jeff and Alicia were there to welcome them. What did it take for them to watch the joyful reunions that they would never share? To hug the men who had hugged their son, who had fought alongside him, who had tried to save him, who had carried Matt on the first steps of the journey that brought him home in a flag-draped coffin a month before?

Henderson would later write in a poem to Matt:

> I must admit, I couldn't say anything, not even 'I'm sorry.'
> Just wept into your Mom's cheek.
> It was her who held me tight, told me it wasn't my fault.

This story began when I asked the members of the RAF Reaper community for their most memorable experiences.

'This was, and still is, the one event which I recall the most. It is never far from my thoughts,' wrote Ross in response. 'Everyone on Reaper will have a particular event that sticks in the mind and this is mine. It was difficult then and it is close by now, but it was illuminating also. Events like this happen; it is what we do. Providing overwatch and support to friendly forces on the ground is challenging but extremely worthwhile. For them to know you are there, silent and unseen, ready to act in their defence, as required, hopefully takes some of the edge off for them. For me it is the most rewarding part of the job.'

He continued, 'I was tempted to make contact with the family but never did. What could I possibly say? I do think of him, especially when 9 June comes around. I look at his memorial webpage from time to time, happy to see how his humour, personality, and positivity are remembered.

Ross's words left me with a dilemma. What should I do with this information?

My first reaction was to set it aside or just give a brief account of an unknown American Marine who died in an incident in a faraway, dusty landscape. Life and history are littered with unheard stories of death and loss in war. But I knew the story.

As I read and re-read Ross's account of what happened, three thoughts emerged. First, this event sums up an important and unappreciated factor at the heart of remote Reaper operations, with intense emotional intimacy across vast distances. Second, all the high technology weaponry in the world cannot stop the unpredictable in a war. And third, Ross was one of a tiny number of people who will have clearly seen the final seconds of Matthew Richard's life and who still remembers them. This story represents the hidden face of war and loss.

I recalled being the chaplain once standing with the Casualty Notification Officer at the end of a garden path, waiting to break the news that no family ever wants to hear. Hoping that my legs would carry me to the front door. Most of all I remembered bereft, grieving family members being desperate for any and every crumb of information about the final hours, minutes and moments of their loved one's life. Somewhere in the US, a family might want to know about those final moments of Matthew Richard's life that Ross witnessed. Knowing about those few seconds would not bring him back, but experience told me the family would likely want to hear about what happened.

I spent a sleepless night agonising over what to do. I decided that it was important for the family to know more, and Ross had given me the option of contacting them. I resolved to do so through the Memorial website.

The website was easy to find and had a helpful 'Contact' button. But how to start the message? What do you write and who might be reading it? Was I about to unleash an emotional tidal wave on a family that had perhaps adapted to their loss and were content with the information they had?

Tentatively, I set out a small amount of information, apologised in advance for any distress and offered to either send more information or to leave the family alone if that was their preference.

Within twenty-four hours the reply came back. It began: 'My name is Jeff. I am Matthew's daddy…' Yes, the family would like to hear more, even though it would be difficult and painful. But most of all they wanted to be put in touch with Ross so that they could thank him for remembering their son.

So I relayed Jeff's message to Ross. Yes, of course he would be happy to be contacted by Matthew's family. He asked for a couple of days to finish his current series of shifts and try to find the right words to capture the memories and feelings that he had carried for six years. What passed between them is private.

From that initial contact came the first snippets of information about Matt – the eulogy from his memorial service, a newspaper report on his death and photos of a young man full of life. I raised the possibility of writing about these events and was put in touch with Seth Hickman, who remains in close contact with the family, as protective of them as he was of Matt. He shared his memories of what happened and introduced several members of the squad. Over several months their fractured recollections provided the first-hand account above.

When I thought this story was complete, Alicia shared the very personal details of what the family went through and sums up the true cost of war:

> The days and years after the initial week have been full of different emotions and changes in our lives. We went to bed on Wednesday 8 June 2011, to wake up to our world completely changed on 9 June. As any grieving parent can

testify, life changes when you lose a child. YOU change. You lose a part of who you are.

As for that question about closure: closure never really comes. There will always be questions, what ifs and whys. Peace, comfort and understanding come slowly with acceptance. Our peace comes in watching the boys of 2/8 – 2nd Battalion, 8th Marine Regiment – live their lives to the fullest; it comes in college graduations, engagements, weddings, new homes, the birth of a child. It comes in hearing one say that he will never squander any opportunity given him because of the opportunity Matthew has given him to live his life. It comes in finding out, six years later, that Matthew is being remembered in a distant land by a drone operator who never met him but was impacted by his death and felt compelled to find out more about him.

It comes in seeing Matthew being honored by our community and those who loved him so well. We now have several children named after our Matthew: Matthew Joseph here in Iota named by a former teacher of Matthew's, a grandson named Matthew James and a Matthew Clark, and Hunter Matthew from squad members. We also continue to honor Matthew through the Cpl. Matthew Richard Memorial Foundation as we sponsor a 5 kilometre race each year, where hundreds of people come out to help us celebrate not only Matthew but all those who have served our country.

Once a year, the squad, and many others, gather in Iota, Louisiana – 'The little town with a big heart' – to run the 5k Memorial Race. Jeff and Alicia have invited Ross to Iota to run with them, and he is making plans to travel out.

They remember what was, and imagine what might have

been. From Louisiana to Los Angeles, the squad has scattered. But they all remain bound by the invisible thread that starts with Matthew Richard and his family, reaches out to his squad members, and leads back to that compound just south of Fort Lambert in Afghanistan. It is a thread that extends across the Atlantic Ocean, where a far-off British airman also remembers.